Chuck Clegg

AF799832

GENESIS

II

Return to the Closed Garden

By

Chuck Clegg

This book is a work of fiction by the author. The characters, names, and events depicted are products of the author's imagination and made into a fictional story. It is purely coincidental if any fictionalized events, organization, characters, or names are similar to actual organizations or people who may be living or dead, now or in the past.

ISBN Paperback - 978-0-9889358-6-0

ISBN EBook -978-0-9889358-7-7

ISBN Hard Copy- 978-0-9889358-8-4

Library of Congress Control Number: 2021920502

Published By: Charles E. Clegg
West Virginia Video Productions
New Martinsville, West Virginia

I would like to thank Tim Haught for allowing me to use a representation of a picture taken by him for use in my book.

Mankind was created with the gift of intellect. This gave him the ability to ask...why and what if. That ability allowed the human race to search for answers in a world of endless questions. Within each of us there are the questions; who are we, how did we begin and could we live forever. For some, there is no need to search for answers. All that is needed to be known is contained in the words and teachings of the Bible. For others the answer is in the search for the knowledge of mankind's creation. Genesis tells us Eve ate forbidden fruit from the tree of knowledge and then gave it to Adam. He then tasted the fruit. Their eyes were opened to consciousness of the world around them. That act set mankind on the path to discovery. On the thirteenth day, Adam's eyes were opened to begin the search for the final truth about the creation of the human race. God set them upon the path from the Garden of Eden and into the world to begin the family of mankind. That path out of Eden has led us to seek answers to one final question, can we live forever? Today we may be at that point in mankind's journey to ask...do we want to know that ultimate answer and its consequences. Perhaps that answer, like the knowledge from the fruit of the tree, will change mankind's path towards the heavens, or towards the abyss of hell. That answer may only be found in mankind's return to the Closed Garden and the rediscovery of Genesis II.

Other books by this author

Return of the Gunboat

Lincolns Gold

Ghost of the River

Summer 2022

Last of the Buffalo Hunters

Dedicated to my wife, Mary

CHAPTER 1

LATITUDE 29.240730 LONGITUDE 49.345093

Doctor Al-Farris sat looking at five large flickering monitor screens on the wall in front of him. His stare was almost frozen from hours of watching the near silent images on the viewing screens. He was sitting in the control room on the 265 foot research ship, *Below the Green Seas*. Beneath the vessel was a small unmanned submersible named, *Searcher IV*. The deep water probe moved slowly along the sea bed of the Persian Gulf. They had completed two-thirds of their patterned search grid.

The near silence of the room was only broken when one of the other three scientists yawned from boredom watching the screens. The room smelled of sweat and an old tuna sandwich lost behind a control panel. The small room was warm and becoming humid from the hot afternoon sun beating down on the metal hull of the ship just above them. At the peak of the day's heat, the onboard air conditioning systems could not eliminate the intense radiant heat on the ship from the sun's relentless searing touch.

Dr. Paul Al-Farris, the lead scientist in searching for the ancient biblical worlds of the Middle East. He dedicated his life to finding lost places spoken of in the Bible. For the last two years he and the other researchers spent time inching their way along the bottom looking for a place long believed lost. It was a place some believed only existed in legends or myth from the Bible, a place known to most religions of mankind. Christianity, Judaism and Islam all tell of this place in their teachings. There are those who believe in a literal interpretation of the Bible and know such a place really did exist. They think otherwise would bring eternal damnation come judgment day. Doctor Al-Farris and his team are believers and now they are trying to find Eden lost beneath the waters of the gulf.

The Bible describes a place known as the "Garden of God" in Genesis. Also in Genesis, God created man whom he placed in the garden, east of Eden. In this place Eve, whom God created from a rib of Adam, tempted him to eat from the tree of knowledge of good and evil. In the passing of time after their banishment from the garden, Eden returned to the embrace of the earth.

Two and a half years ago a NASA satellite began mapping the area and detected two ancient dry river beds. The Bible tells of four rivers that flowed into the garden. Al-Farris believed those rivers to be the Pishon and Gihon, long dead rivers. But, the Tigris and Euphrates still flow across the region.

Biblical scholars for centuries have sought to find the birthplace of men, according to the words in the Bible. To find such a place would confirm that God is real and not a mere metaphor for the beginning of man in a religion-based fairy tale. Non-believers have clamored for real proof throughout the centuries. Al-Farris and his team hoped that the answers were below them, hidden deep in the sand and mud of the bottom.

On the sea floor, the *Searcher IV* has been working a grid pattern in the area where Al-Farris believed the garden laid beneath the bottom. The doctor believed the evidence of the four rivers convergence is a strong indication of the gardens location. In the time of Adam, the once fertile green land was above the surface of the water in the delta flats of four rivers. Over thousands of years ocean levels increased and the once fresh water lake and fertile lands of Eden were flooded by the salt waters of the Persian Gulf.

Below the ship, an unmanned submersible worked an invisible grid pattern across the ancient sea bed. At precise points in its programed search pattern, it would settle to the bottom and insert Deep Sonic Thumpers. Each of these devices will on command from the surface send out long angle sound waves down into the bottom mud. These types of sound waves have been used to communicate with submarines over long distances in the vast depths of the ocean. The technology has been developed in recent years to map structures below the earth's surface. It is commonly used to map faults in the earth's crust, along fault lines. With the 3-D mapping technology, small movements in faults can be mapped and compared to previous images. The data can then be analyzed and then a determination can be made if a dangerous situation is present below the earth's surface. The submersible also sends out a series of subsonic energy waves and microwaves. Those electronic waves also travel deep into the soft bottom. If they come in contact with anything buried in the mud, the signal will bounce back to a receiver on the bottom of the boat. It could detect solid objects several meters deep into the mud. Computers onboard the ship collect the returning data to the submersible along with the ship's data collection and then combine the information and analyze it. Once analyzed it can build a remarkably clear image in 3-D of the bottom mud down to a depth of two hundred meters.

In the two years since they began working the search pattern, very little of any consequences had been reflected back to the flickering screens. Occasionally an old car or sunken ships have been found. Once they detected the image of an airplane. Most likely it was lost during the war.

In the quiet room a young research assistant asked Dr. Al-Farris, "Sir, would you like a cup of tea?"

He stretched his arms over his head as he responded, "Yes, I would be very happy to accept your offer. And do we have any of those crackers you brought from home?"

The young assistant indicated with a nod of her head there were some left. As she made his tea, the doctor leaned back in his chair and rubbed his tired eyes. Then he said, "My friends, when I finish this delightful cup of tea, I am going to retire for a nap. I will come back this evening. I am beginning to believe….."

At that moment the screen began showing a ghostly image. At first it was just a bunch of squiggly lines on the screen just enough at first to get the room's attention. But, as quickly as it appeared it faded. A collective sigh could be heard in the room at the sudden change of events.

Then the image reappeared in much sharper detail. Al-Farris told the operator to place the images on all the monitoring screens in the room. The doctor called out to his assistant who was now reading the data coming into the computer. "How deep are those images coming from?"

The young man hurriedly calculated the depth, as Al-Farris kept impatiently asking, "How deep?"

Finally the young man said, "One-hundred and seventy-five meters sir."

Al-Farris slowly walked to the main viewing monitor and touched the electronic screen. He acted as if he could feel the images as they moved across the screen. "Praise be to God…could it be we have truly found it."

One of the others on the research team repeated, "Praise be to God!"

He turned to the team of scientists who were now wide awake, watching in wide eyed fascination. Each of them was wondering in their own mind, "Was this the place that Genesis told of?"

As he watched he looked at the position indicator for the ship. It shows Latitude 29.240730, Longitude 49.345093. He pointed toward the red flickering numbers. "Set a marker on the navigation chart so we don't lose our position. And bring up Searcher. We'll send down the ROV drill platform. We'll see what it brings back to us."

The remotely operated vehicle was a specially built automated drilling unit that could drill to a depth of four hundred meters.

Al-Farris turned to the people in the room who were still mesmerized by the ghostly images. "People, we have a lot of work to do. We do not know exactly what we have found. Let's not jump to any conclusions. We do this step by step to be sure we don't make any mistakes. This may be the greatest discovery of our lifetime or simply an abnormality in the deep mud. Either way, let's not mess it up. The ROV's sample drill will hopefully bring back some evidence. We need to be sure what's down there. Now, let's all do our jobs and get the ROV lowered and retrieve a sample."

It took nearly seven hours to ready the deep drilling ROV unit to reach the bottom and position itself. The location information from the searcher probe was downloaded into the automated drilling unit's onboard guidance system. Once programed it could pinpoint its target to within a one square meter location.

On the bottom, it would use the searcher's readings along with onboard sensors to compute the proper drilling speed and angle. There was an excitement in the room when the ROV unit finally began its search. The team could see the progress of the drilling unit into the bottom on the room's computer monitors. Slowly the telescopic drill penetrated the bottom mud in search of long awaited answers.

The drill unit had been developed for testing bottom samples before construction of underwater bridge pilings and oil platforms began. In the center of the drill head is a special chamber that was designed to be remotely opened to sample bottom material. Unfortunately, it was too slow for construction site samples, when hundreds may be needed. The unit was sold to a university in the Middle East for deep water research. The drilling platform carried with it onboard, 120 sections of drill pipe. As each 20 meter section of pipe was drilled into the bottom, an automated arm would attach another section of pipe. The unit could reach depths of 400 meters with pinpoint accuracy. So far, the only thing of interest it had found were bits of pottery and fire ash from ancient campsites.

The first six months of the mission they drilled countless test holes. Those core samples were analyzed on board to establish the age of bottom sediment. It was determined they needed to sample between 150 and 200 meters.

Hours passed and the clarity of the water became murky, as the drill penetrated deeper into the bottom. Clouds of silt now obscures the view on the control room screens from the scientist's eyes. Finally, the technician monitoring the drills operations called out. "The torque on the drill head changed suddenly. The drill head sensor indicates it's possibly organic material."

He turned to Al-Farris, "Do you want me to open the sample chamber?" The doctor nodded his head to begin.

No one said a word as samples were being taken far below in the mud. Once the samples were taken, the drilling operation went into reverse. Slowly the remote drill unit pulled the piping and placed it back into its holding racks. It took nearly ten hours for the unit to finish its task on the bottom. On the ship's back deck, scientists stood waiting, staring at the heavy cable as it slowly pulled the drilling ROV toward the surface. Each person was hoping that the answer would be coming to the surface. Answers that could change the world. Answers that could redefined what we believe.

Slowly the heavy mechanical rover was placed back on the deck. The research team calmly helped the doctor remove the metal sample canister. The special drill head had worked perfectly retrieving the samples. Anticipation grew as they waited to see what the drill had brought back from deep into the bottom mud. Finally, the metal tube was laid carefully on the table. Doctor Al-Farris began removing the tube's seal cap. He stopped abruptly and asked his young assistant with a camera by her side, "You are recording this, aren't you?" His assistant, who was so engrossed in the possible historic event, had become mesmerized looking at the tube and she had forgotten to begin recording the event. Quickly she turned on the camera and in a moment she said, "I am recording now, doctor." Once again Al-Farris began carefully removing the containment caps. Everyone moved closer to the table, hoping to be the first to see what was inside. With the end caps removed, he gently tipped the specimen tube and its contents slid onto the stainless table with a soft thud. As it did, a collective sigh was heard as someone said, "Oh God, can it be true?"

Chapter 2
Secrets of the Smoke Hole

Charlie Morgan sat quietly drinking coffee in his small camp along the South Branch of the Potomac River. His camp consisted of a blue tent in the shape of a dome, a darkened ring of rocks which contained his campfire and in a small nearby tree a make shift clothe line. The line was used to dry his fishing boots and tie his food up away from small black ants that made a nuisance of themselves around his camp. It was a hidden place about as far from civilization as could be found in the mountains of West Virginia. It was a place filled with a lifetime of memories pursuing the only thing he had left that he loved, fishing.

He first came to this place in 1955 with his dad. At age seven it was hard for him to follow his dad over the inhospitable terrain along the Smoke Hole Canyon. It was a place still so remote that nature seemed untouched yet by man. The old canyon lay in the eastern part of the state, not far from the Maryland state line as the crow flies. That long ago time, when he came to this special place with his dad, Charlie never complained. With his short legs he struggled to carry his fishing pole and keep up on the rocky trail.

The Smoke Hole was cut into the old rocks that formed the steep canyon walls in a time long before man. For thousands of years flowing waters from retreating glaciers eroded the hard stoney earth and carried it toward the flat lands. In this place, time had no meaning. The steep cliffs reflected only the sounds of soaring eagles and hawks, no sound of cars or man's clanking machinery. As far back as history had been recorded, it has always been a 20 mile stretch of pristine cold mountain water. Hiding in the dark swift water was the most elusive of all fish, trout.

A great deal of the river was only accessible by footpath or small boat. The best places were sought out by only those who trekked along the steep canyon and sought the secret fishing holes. Places even today; only a few have listened to nature with man nowhere to be heard. These are deep, cold, and dark waters that give cover to the smartest of all trout that live in such places. At least that is the way avid trout fisherman saw the pursuit of the elusive canyon fish. A place beyond the boundaries of tourist fishermen with plastic reels and fancy fish finders with GPS locators.

As Charlie sat watching smoke from his fire drift lazily up towards the coming day's light, he remembers that hundreds of years ago Native Americans used the many caves in the canyon to smoke fish and wild game. He told himself the fish he caught were direct descendants of the fish that long ago lived in the same waters. They were pure in breed. The only changes were the ones Mother Nature intended to make with the passing of time. Fish that were strong survived in this place. The weak died quickly and did not perpetuate a weakened heritage.

The smoky haze from his fire hung over the ancient canyon, filling the air with the sharp odor of the burning wood. Charlie occasionally tipped his head to prevent being choked by the smoke. Invisible air currents were constantly changing the direction of the bluish smoke from his smoldering fire. His eyes often burned as he closed them, while trying not to breathe until the smoke moved away.

His camp was small, and Charlie made no effort to make it more comfortable. He wanted to be close to nature and not have the trappings of civilization getting in the way of his time in the canyon. Charlie returned to this place at least once a year for as long as he could remember. It was a place that seems to clear his mind of the problems he faced daily outside the canyon. It was a world Charlie had come to hate for what it had become. He found no redeeming qualities to civilization. And in this harsh place, civilization did not venture. The weak could not and did not enter this place of solitude.

Those weaklings who embraced civilization would have no cell phone reception. Their coffee would not be served in a cup with a fancy wrapper designed to not burn their hands as they held it. Charlie enjoyed holding the hot cup between his hands. Feeling the sting from the heat told him he was alive. For some reason, he took a sense of pleasure in drinking from a once shiny metal cup that had never seen the inside of a dishwasher. Fact is, it had never been washed out properly. Charlie rinsed it in the flowing waters of the cold stream. He figured if the water is clean enough for trout, it was clean enough to rinse his coffee cup.

The canyon held nothing for those who loved the comforts of civilization. For Charlie it held everything he came to value in his life, serenity with his thoughts and nature.

After his wife, Rose and son, Timmy were killed by a drunk driver named Joseph Chadwick, he wondered if living in a world without them was possible. He had a hard time getting over the fact they died without having said goodbye. They died for no good reason except the failings from another human being. Chadwick, who killed them had been arrested four times for DUI. And because his family had money, he was prevented from going to jail on small technicalities in the law. That was until he killed Charlie's family. Chadwick is now serving at least eight years in a private prison. His family wanted to keep him safe from the real criminal elements in a state prison. The family's four lawyers made a compelling case that the man had a terrible problem, and he was remorseful for his crime. Charlie figured he was just a spoiled rich criminal who always had money to bail him out of problems. Charlie knew someday Chadwick was going to get his justice, but still, it would not bring back his wife and son.

Charlie's lawyer made sure evidence was entered into the case showing his wife's income per year. He knew it would help in the civil action he filed when the man was convicted. He managed to get the trial held in a county where seventy percent of the residents were considered church members. The lawyer presented to the jury pictures of the terrible accident site. He revealed pictures of drunken Chadwick at the police station when he was arrested after the accident. Chadwick grinned at the camera while being booked for the lethal accident. Charlie's lawyer paused as he took a moment to look each juror in the eye. He pointed towards Chadwick as he said in a cold voice to the jury. "Joseph Chadwick sat there grinning for his picture, at the same time Rose and young Timmy's bodies grew cold in the morgue." The jury was out for 20 minutes, they found Chadwick guilty of two counts of vehicular manslaughter. Being the fourth time he was arrested for DUI, it also was stipulated to be a felony.

Later Charlie filed a civil suit and asked for two million dollars in lost earnings for his wife who was a top doctor in her field and taught at the medical college. The jury granted the award and also awarded him 10 million in punitive damages. Charlie knew Chadwick nor his family would ever pay a penny, but with the judgement hanging over them, they would be in financial limbo for many years. The drunken bastard is still locked up and he owes 12 million dollars. Charlie knows that Chadwick will never pay, but neither will he ever have a penny to his name. His family will have to support him with underhanded money out the back door for as long as he lives. If Chadwick continues to drink after getting out of jail in eight years, that most likely won't be long.

The money meant nothing to Charlie. It was not about the money that drove him to lock Joseph Chadwick up forever if he could. No amount of money could fill the emptiness left by the loss of his wife and son. Since Rose and Timmy died, Charlie had forced himself into his work, not taking any time for holidays or to share a meal or laughter with friends. He did not even want to just sit on the front porch and watch the day go by, something Rose and he liked to do after a long day in the summer. They would sit and watch Timmy play in the front yard. They sometimes talked of her flowers or his idea for a new trout fly. They hoped their son would follow Charlie's footsteps and attend Tygart University. Whatever Timmy decided to do with his life, they would be proud of him. Those conversations meant nothing now. Those words Charlie and Rose spoke to each other meant nothing to any that passed by their porch in those distant days. But, to him they were precious times together when they shared love and dreams. Now, the memory of those times on the porch just reminded him of his pain. He missed just sitting and talking with Rose.

His occasional trips into the canyon were the only times he and Rose were apart since they were married in 1969. Rose liked the comforts of her world and the benefits of civilization, a warm bed, hot showers, and Days of Our lives on TV. The canyon where Charlie camped and fished provided none of those things. But, she realized he found strength in the canyon filled with smoky haze. She knew his quest to catch a record trout on one of his handmade flies was part of his dream. It was a place where Charlie Morgan found solace in life. When Rose and Timmy were alive, Charlie couldn't wait until Timmy was old enough to join him in the ancient canyon. He then could pass on the art of fishing for the elusive trout, just like his dad had taught him. In a moment on a stretch of dark road, a worthless bastard took all those hopes and dreams.

Charlie remembers one of his trips to the canyon, while Rose was still alive. He was setting up his tent, in the same place he and his dad had made their camps. As he unrolled his sleeping bag, he found inside a small bag of dark chocolate covered coffee beans. Along with the bag of beans he found a handwritten note from Rose. It said, "Good luck in catching The General. I figure if all else fails, just offer him a chocolate bean. They always made you open your mouth for just one more. Love Rose and Timmy." Charlie carries that note with him to this day. When he feels the terrible emptiness within the darkness of his soul and wonders, "why go on", he removes it from his wallet and mutters to himself the words written on the worn piece of paper. Longingly he looks at their picture. For just a moment he can almost hear Rose's voice. Then it fades into his loneliness. Charlie then reminds himself that Rose and Timmy would want him to go on. Go on and catch "The General." And when that day comes, he will hold the fish high and call to the heavens, "I did it."

Rose's reference to "The General" was to a legendary rainbow trout that lived in a dark hole of water, just below a long quarter mile riffle. The riffle constantly brought down hapless insects that had fallen into the quick moving waters and then flowed into the deep slow-moving pool below. Below the surface waited "The General," waiting to take the next meal that came into his sight on the waters swift cold current.

The trick to catching the General was to figure out what the mayfly hatch was going to be that day or even that hour. He was a wise old fish that waited with unending patience for the next offering.

Charlie called him "The General" because he almost caught him one time. He had rolled his line across the water in a perfect cast to the far side of a rock boulder that lay hidden below the water's surface. He figured on the far side in the shadow of the stone, a smart trout may be waiting for his meal. That morning the water's surface was covered in drifting flotsam from the Cotton Wood trees along the creek. The floating balls of fluff made it difficult to work his fly across the surface, keeping a natural appearance to the fly's movement.

As he worked his fly through the cotton wood seeds, he saw a flash of red and silver below his fly. Suddenly the water exploded with a strike on his offering. For just a split second he could see the eyes of the trout as it took the fly into its mouth and disappeared back below the surface. In another moment, the trout reappeared and leaped high out of the water in an attempt to shake the hook from his mouth. In that split second Charlie saw something in the color pattern on the side of the fish. The white spots gave him the impression of the pattern on a two-star general's flag. Charlie knew that was not possible, but still he decided to call the fish, "The General". Charlie figured "The General" was just a fish and he was neither an officer or enlisted, he simply wanted his next meal. So far "The General" had not taken the bait or come to the surface to say hello.

These days, Charlie just wanted to feel anything that gave him some comfort. His annual pursuit of a fish with a name in some small way filled that measure of need, but only while he was alone in the canyon of smoke and memories.

Sparks leapt from the burning logs and smoke once again swirled into Charlie's eyes. He covered his face with his hands to prevent the smoke getting into his eyes. As he opened them, they started to burn from the effects of the smoke and tears running down his face. He breathed deep fresh air as the smoke moved away and then finished the last of his coffee. He shook his cup toward the fire to empty the stray coffee grounds in the bottom of his tin cup. As the coffee droplets touched the burning wood it made a quick hiss as if a snake was hidden in the flames. It reminded Charlie of the many rattlesnakes that waited in the canyon among the rocks. He had come close several times to being bitten while climbing along the edge of the canyon as he surprised a sleeping snake. Fortunately, those early morning encounters found the snake still cold and stiff from the cool temperatures of the mountains. The surprised snake wanted no part of Charlie and slowly moved away to wait for the warmth of the sun when it entered the deep canyon.

Charlie looked down the canyon and noticed the fog was lifting. It rose toward the coming daylight, still a couple of hours away. The steep canyon walls would not let the light in until the sun was nearly overhead. Charlie knew it was time to head off toward his encounter with "The General." Somehow in his mind he knew this may be one of the last times he and his old friend would pit their skills against each other in this ancient place. He would try to catch the elusive fish and "The General" would pretend to be interested in taking his offer on the end of his fly line.

Charlie had pondered even coming back to this ancient canyon this year. His knees were growing more unsteady with each passing month. Sometimes he felt they were not going to support him for another day, especially when it was cold and raining. His doctor explained he could have them operated on and afterward they should be better, or at least, not so much pain. Somehow after Rose and Timmy's death, he could not bring himself to think of anything that would lessen his own body's pain. He realized an operation was no guarantee that he would be able to track along the rocky canyon trails and visit the waters below the long riffle. He felt the last thing in his life that mattered was slipping away, like the blue smoke from his campfire.

As he drove to the canyon this year, he remembered seeing places along the swift moving creek where the Division of Natural Resources had built handicapped accessible fishing places. Concrete ramps ran down to wooden platforms overlooking the stream. A place that was designed to give people with disabilities a chance to catch fish. Below the handicapped platforms the DNR bulldozed deep holes into the creek's bottom. Each day during the warm season a truck from the Bowden Hatchery would stop. A hatchery worker would step out and go to the back of the truck and dip a net full of wriggling fish. They would then hurry down onto the overlook and quickly dump the fish into the cold water below. The newly released fish quickly disappeared into the dark cold water.

Most of the time no one with a disability was there to cast a line and try and catch the newly released fish. But, whether those with disabilities were there or not, the so called road fisherman who followed stock trucks swarmed to where the newly stocked fish were released. These so called sportsman quickly cleared the waters of the trout. Charlie wondered how those road fishermen called themselves sportsmen.

He remembers seeing the stock truck moving slowly along the road that follows the creek. Behind the truck was a caravan of trucks and cars waiting to see where fish would be released. As quickly as they were released into the streams, road fisherman sprang from their vehicles casting their lines into the water. It was like catching fish in a barrel.

He knew sometimes one hatchery truck would leave moving slowly along the creek roads. It was a decoy used to draw road fisherman along for a slow trip toward the Virginia border. When the caravan of road fisherman disappeared in pursuit of the decoy truck, a second was sent out to deliver fish into streams around the state.

Charlie never figured that was fishing. Neither could he see himself sitting in a wheelchair trying to catch a fish that a man had dropped into the water beside his bait. Besides, those trout were raised in concrete hatchery pools. The fish did not hunt their food in those places. They waited below automatic feeders filled with something that looked like dog food. He once thought to himself, "Hell, if I was go'na fish like that, I'd toss in an M-80 and bring them all to the surface. They would not prosecute me; I'd be handicapped, and I have my rights."

Charlie, like his father before him, believed the only real fish were native trout and you had to pursue them on their terms in the wild. If you could hear cars passing or kids laughing, you were to close to civilization. You had to be in a place where you and nature shared the sounds of the river as it passed through the steep walls of the Smoke Hole Canyon. In this place, native fish could still reproduce on their own. Their eggs weren't squeezed out of the belly of a brood trout by a hatchery worker and then mixed with sperm from a fish that never touched the gravel bottom of a stream. No, the trout in this part of the canyon were real fish whose ancestors were the same ones caught and dried for food in the caves of the canyon by real Native Americans.

Charlie believed real fish were when two fish fell in love one spring morning in the swift cold waters and made love for just a moment. Eggs were spawned by a female into the gravel of the bottom and then the male fertilized them with his sperm. With a little luck and by choice of nature the strongest of the eggs hatched and struggled to live and grow into real fish, not manufactured representations of what man created without much help from nature. Charlie figured there was just something wrong with science creating anything alive, especially a fish.

As the light of day first touched the top of the mountain, Charlie had begun making his way along the steep narrow path that led farther into the canyon. His fishing equipment was secured into his backpack. This allowed him to navigate with two walking sticks. They helped to steady his steps along the narrow path. On rare occasions he would use them to move a copperhead or rattlesnake off the rocky path. He saw no need to kill them. He figured if some tender foot fisherman were to follow him into the canyon and ran across the same snake, perhaps they would be deterred and turn back and fish the handicapped area. Besides, most of the snakes he encountered in the canyon were willing to move aside and allow him to pass. Snakes he met outside the canyon were always ready to bite. Those kinds of snakes were the two-legged kind.

With the path guarded by snakes, most tourist fisherman did not venture this far into the canyon. Last thing he wanted to do when fishing was to answer the same old question, "You catchin' anything?"

He kept a small jar of pink marshmallows in his vest pocket to show any inquisitive fisherman the bait of the day. He even took a magic marker and made smiley faces on the sides of the marshmallows to show to those who came asking, "What are they biting on?" And if he got hungry, the smiling marshmallows could be eaten as a quick snack. Fortunately, most fisherman he encountered this far into the canyon were like him, looking to be alone with nature. They did not ask or offer suggestions. Most simply tipped their head and offered a "howdy" as they passed. It was an unwritten rule if you saw another fisherman, you remembered the time they passed. Charlie would begrudgingly return the gesture with a tip of his head. Each real fisherman knew no fish were caught while polite conversation was being made and time lost talking, was time lost fishing.

At least once in the past forty years a fisherman did not return come the end of the day. At the local fishing lodge, conservation officers asked if anyone had seen the missing man. Those that had seen him could recall the place and time of their meeting. This information helped officers to have an idea how far the lost man may have ventured into the canyon. In that case the missing fisherman had fallen and broken his leg on a slide of lose shale that had fallen from the canyon wall. The angler was unhappy that he had also broken his favorite bamboo fishing rod. The fact was, he was mostly upset about the fishing pole. A broken leg can be repaired, but a broken handmade bamboo fishing pole… it will never have the same action or feel of the fly after it has been repaired.

It took Charlie nearly an hour and a half to work his way along the narrow path deeper into the canyon to where his favorite spot waited for him. It looked different these days than it did when he came here as a young boy with his dad. Back then trout fishing had a season. Each year for weeks ahead of time his dad would check his gear to make sure it was ready for the first day. Flies and lures were inspected with a magnifying glass to make sure the feathers and hairs were exactly in place. If any fly showed damage or a bit of wear, a replacement was made. Charlie remembered sitting watching his dad position the hook in the holder and select the correct material for the body. It may have been a bit of yarn or a piece of colored foam. He would then trim a hackle of hair or a feather to be placed just right as he began slowly moving the thread spool around the body to secure the fly together. As he did he would tell Charlie about the big trout he had caught over his lifetime. He used to laugh and say, "I caught them real trout, not them damned goldfish trout from concrete hatchery ponds. Yes sir, they were natural fish, Brookies, Browns, Rainbows. Fish that their first gulp of water ran down from the tops of them West Virginia hills. God's fish, in God's creeks." Charlie would listen for hours as his dad tied flies and told stories of the old days. He suspected some were told just for his benefit, but it did not matter, he loved to hear his dad tell them. That was a lifetime ago. Charlie now had no son to tell his stories of trying to catch The General. Sometimes he thought he was so grouchy because his fish stories were all built up inside him. He also knew his stories and those of his dad were going to die with him someday. He had no son or grandchildren like God intended for a man to leave his favorite fishing pole and stories of the big one that didn't get away. Charlie was just an old man who had lived too long after his wife had died. His secretary once told him to write a book and tell the stories of the years he and his dad had spent fishing the Smoke Hole. Maybe that way they would be remembered by someone and it damn well could not hurt to make Charlie less grouchy.

Forty years ago, the area experienced what some called, "A thousand year flood." Heavy rains came in the dark of the night and when daylight returned, homes, cars and lives were lost to the rushing flood waters trapped between the hills of the valley.

Streams with large stones that had been there for as long as men could remember were washed away in the aftermath of the flood. The first time Charlie returned the following spring, he could not recognize the place he had known so well. Nature and the Division of Natural Resources had begun working to restore the area to a semblance of what it was before the terrible flood. The creek could be repaired and someday would seem normal to those who had not known it before the great flood. But no amount of work could change the aftermath of lost lives that ended in the flood. For several years it just did not seem the same.

The place Charlie liked to fish for the General was a deep hole of water below a long riffle. The creek made a hard turn just below the shallow water and cut deeply into the hillside, creating a deep hole of water with a rock shelf on the far side. In the middle, a large rock the size of a pick-up truck was submerged just below the water's surface. Both the rock shelf and the bolder held the potential to hide a big fish.

Charlie sat down and removed his backpack. He felt inside and found a bottle of water. It was still cool in the canyon and the sun was just minutes away from making its daily appearance from above. The long walk had brought on a sweat under his shirt. He hesitated to open it and let the cool air into his skin. He just sat and drank the water, looking at the stream while formulating a plan of attack to entice "The General" out of his hiding place.

"All right you old bastard, where are you this year? I know you like to hide under the rock shelf if you have eaten. But I'm guessing you're waiting for the days hatch to come up with the sun."

Charlie knew when the sun touched the water, a hatch of small insects would take place. As quickly as they flew into the sky, some would falter and fall back onto the water. Their tiny bodies barely make a dent in the water's surface almost unseen to most eyes. But, below the surface in the shadows, hungry trout did not miss anything that touched the water as they waited for the first meal of the day. Charlie's first challenge was to figure out if the always wary trout were interested in food that floated along the surface or sank slowly toward the waiting fish.

The choosing of a wet or a dry fly could be the difference between catching "The General" or just wetting his line.

He sat studying as bits of flotsam floated down the stream. As they moved over the swift waters of the riffle, they swirled toward the far rock shelf. Fish are smart. No sense in chasing breakfast if it will come to you. He knew what he was going to do.

Charlie removed his custom-made bamboo fly rod he had secured to his backpack before leaving camp that morning. Carefully he assembled his pole and threaded the line through the pole's guide eyes. He tied a leader to the line. Next, he tied on a dark tippet line. He liked the dark because he thought it had a little more weight to it. That helped to give him the casting distance to reach the far side of the creek. Finally, he opened the small leather pouch Rose had given him on their 30th anniversary. It contained an assortment of handmade fishing flies. He thought for just a moment. "Light or dark?" After looking over the selection of flies available to him he decided. The sky was bright blue with the coming sun. A light green hackled fly would be just the thing to entice "The General" out from under the rock shelf if he was there.

Finally he was ready. With the sun coming from behind him, he would cast a long shadow the moment the sun broke over the ridge. He bent low to lessen his shadow on the water's surface.

He managed to get to a spot just at the lower end of the riffle's swift moving waters. He hoped the bubbles and turbulence on the surface would mask his silhouette from the fish below. It was time to fish.

He stripped his line and began working it back and forth to get just the right distance. Finally, he had all the line moving as he pointed his pole toward the far bank. It landed gently on the water's surface where the current moved toward the bank. It began moving along toward the dark water just off the rock shelf.

He watched and waited for "The General's" appearance. It was still and he saw no movement in the clear water. Just as he was going to pull back his line for another cast, he saw a quick darting movement below the surface. Swirling water near his fly told him something was interested. He knew a trout had come out to take a look. The fish did not take the offering this time.

Charlie knew the fish would have taken it if it wanted the color. He made a quick decision, switch to a mayfly hatch. Small and golden in color. He had seen a couple of dead mayflies along the bank last evening. Perhaps the fish still were interested in the color that was in the water yesterday. He opened the pouch and found a small gold colored mayfly lure. It was a little larger than what he had seen along the stream, but he hoped the size would make it more attractive to a hungry trout. Quickly he tied it on and again pulls 30 feet of line out.

He carefully moved the flexible bamboo pole back and forth, getting his line to just the right spot as he pointed his poles tip and allowed the fly to land on the water's surface. He bent down and squinted his eyes as he watched for the fish to come out and take his fly.

Then a shadow moved quickly from the dark depths. It stopped and hung just below the floating fly. Just then the sun crested the ridge behind him, and a ray of sunlight penetrated the water just where the fish was holding steady.

"It's him", Charlie muttered to himself. That distinctive color pattern of the fish in the clear water glowed in the first rays of the morning sun. Charlie felt a rush of anticipation as he waited for the fish to make his move.

"Come on, take the damn bait," he said in hushed tone to himself.

"The General" began to rise, moving slowly toward the drifting fly. Charlie gave it one twitch, just enough to give the appearance it was alive.

Closer the great rainbow trout moved toward the fly, then suddenly from nowhere a flash of gold. From the depths, a 20-inch golden trout came up and had taken the fly right out from the nose of The General.

The golden trout put up a great fight as it made a quick downstream run trying to find swift water in an effort to free itself. It jumped three times, hoping to throw the fly from its mouth. Instinctively Charlie knew from his years of experience every move to prevent the fish from getting away. He held his pole high, leading the tiring fish toward the near shore. Carefully he moved along the bank to a place where the water shallowed. He knew if he could get the fish to that place he would have control. After a couple more minutes the tired fish gave in. He knelt down and slid his hand under the middle of the magnificent golden trout.

He held the fish and looked at it for a few moments. Then he raised his head and looked around the canyon as if to see if anyone had seen him catch the beautiful fish. Then he looked down at the fish he held in his hand and explained loudly, "You bastard, I was about to catch 'The General', and you had to come up and steal the fly." He pulled the hook from the corner of the fish's jaw. He raised the fish high and slammed it down against a rock. The force was so great the fish did not flop around on the ground. It only had enough life lift in its body to gasp for a few more breaths of air. Its eyes became fixed and lifeless as it breathed its last. The fish was dead.

"Damned genetic half breed. God did not make a golden trout. Some damn fish biologist decided to breed an abnormal fish to create a new species of goldfish. Probably just wanted to get his name in some book as the creator of a new species, playin' God. Well, this is one less of them damned mutants.

Besides, I don't have to worry about them spawning. They are incapable of spawning in the wild. If God had intended them to be in the water, he would have created them himself. Man should leave nature alone. Maybe this fish would have been a good goldfish for some kid's glass bowl in his bedroom, but not in the native streams of the canyon."

He took his foot and kicked the fish several times working it back toward the water. Finally, at the water's edge he kicked it away from the shore. He watched as the once great golden trout slowly sank to the bottom. "Well, at least it will be good for something...crawdad food."

Charlie had always disliked the golden trout. He caught his first one in the summer of 1963. Figured it was some weird albino colored fish. At first, he did not even want to handle the fish at all. He had watched too many monster movies in the late 1950s with strange radioactive creatures trying to take over the world. That weird golden trout was a glowing mutation of a creature from the darkened depths of the canyon. His dad had heard the state bred a rainbow trout that had a golden coloration first discovered in 1955. Biologist's worked with the golden strain until it was perfected in the early sixties. Most likely some politician decided that with the state's golden centennial in 63, the trout could be given a name to commemorate the event. It was named, Centennial Golden Trout.

Charlie and his dad just figured they were a damn mutant fish. The bright color would make them easy prey for predators in the wild and would not last in the mountain streams. That was over fifty years ago. The golden trout was still alive and prized as a trophy fish by most people, but not Charlie. He made a point out of killing every mutant fish he caught. And today after this golden trout caused Charlie to miss what may be his last chance to catch The General, it had to pay with its life. It now lay on the stream bottom being feasted on by crawdads.

Charlie retied his line and worked his way along the bank to make another attempt to lure "The General" out from his hiding place. Cast after cast and no offer to take his best fly.

Finally, after an hour, a small rock bass took his fly, swallowing it down to its gizzard. Charlie knew the sun was too high and the fish would not feed until later in the day. He stood looking at the waters flowing by and thought about "The General" and how much he loved fishing. He never gave a thought to the life of the golden trout. It had no meaning. It was not even worth remembering as he enjoyed the moment in the serenity of the canyon.

As he sat, he sensed someone watching him. Not sure how he knew someone was there, but he carefully turned to look toward the path along the edge of the creek.

Above on the high trail that back packers used, he saw the silhouette of two people looking down at him. The sun was high above, making it difficult to make out who the two were. Charlie shaded his eyes from the bright sun as he tried to make out who was watching him. Finally, his eyes adjusted enough that he could see each wore clothes not for hiking in the canyon.

He heard one of them call to him; he could not make out what they were saying over the sounds of the creek echoing off the steep rock walls. He held his hand near his ear indicating to the man he could not hear his words.

Finally he heard the man's words, "ARE YOU CHARLES MORGAN?"

"Shit," he thought to himself, what the hell do they want? He quickly thought of the fish he'd killed that morning. Maybe some damn tree hugger had seen him and turned him in. He knew this stretch of water was catch and release. That meant the fish were to be released alive unhurt. Some concept to make fishing more enjoyable for tourists. Charlie just figured it was another way the government was trying to interfere with people's lives. If that's what they wanted, he would have to talk his way out of this situation, something he had done before.

He called up to the people standing high above on the canyon rim. "YEA, I'M CHARLIE MORGAN, WHAT DO YOU WANT?"

The two did not answer his inquiry. They just started making their way down to the edge of the creek where Charlie was standing. The way they were dressed and carried themselves told Charlie they were most likely the law. In fact he knew they were most likely U.S. Marshals by their appearance.

Charlie was a little concerned. Why would two federal enforcement officers come after him this far in the canyon? A murdered trout was no reason to send out the law in force. Besides unless one of these two intends to go swimming to find the evidence at the bottom of the stream, it would be his word against some damned tree hugger's complaint.

As the two made their way closer, Charlie got another surprise. One of the two was a woman.

She spoke first, "you Charles Morgan? Retired federal prosecutor for the Justice Department?"

The question changed Charlie's demeanor instantly. He was no longer a guy fishing along a stream. He was being addressed as a federal prosecutor. "Yes, that's who I am, and can I inquire as to your interest in me, especially way out here? I left no word of my location, and frankly did not want my whereabouts known for obvious reasons. My secretary was not to give out my location to anyone."

"Sir, I am Marshall Harris and my associate is Marshall Hickman. Mr. Morgan, Judge Peterson dispatched us to find you. He sent us into this canyon. He also told us to ask nicely for you to return with us to Elkins. He said ..." The young woman marshal paused for a moment.

"Well, what did Peterson say?"

"Well sir, he said, quote," "Go get that grouch and tell him to get his ass back here today. And if he won't come, use your stun gun and drag him." The two marshals glanced at each other and tried to conceal smiles.

"So, Stinky Peterson said stun me did he?"

Charlie leaned towards the two marshals and grinned, "You know why I call him stinky?"

The two marshals tried to maintain proper manners, stood straight and made no indication they wanted to know how his nick name came about.

Charlie laughed, "Since you're wanting to know and being on duty won't let you ask, I'll tell you anyhow. Not long after he was named to be a federal judge, he was presiding over a case that had gone on for a month. Important case involving a mining company who was being sued by the government. The lawyer for the coal company was making his closing argument. Meticulously working through every fine point of information he wanted the jury to remember before they began their deliberations. Peterson, not wanting to interrupt the man, felt the sudden call of nature. Two cups of coffee and a hot brand muffin that morning wanted to move things along. He held it as long as he could, not wanting to interrupt the lawyer's final words to the jury. Finally, he could hold it no longer. His sphincter muscle gave way and he passed a silent gas cloud into his courtroom. Covered in his black robe and high on the bench, he hoped no one took notice. But, as the company attorney was finishing his presentation, he began noticing a foul smell. For a moment he sniffed the air like a rabbit sensing danger. As he turned to the bench he noticed the judge had an odd look on his face. The attorney wasn't sure who had passed the gas, but he quickly concluded his comments and asked the judge for a short recess. Peterson reluctantly agreed, acting as if he wanted to continue and conclude the two sides closing remarks before turning it over to the jury. His reluctance did not fool anyone, as the odor settled across the front of the courtroom. Since then, them that know the facts of his first case still like to kid him about his experience with the unforeseen circumstances during a trial. He denies it to this day."

The two marshals could no longer contain their amusement as they broke out laughing for a moment. After a few seconds, they contained themselves to proper manners for federal law enforcement officers. The young lady looked at the other marshal and they once again broke out in momentary laughter.

Finally smiling as she said, "Mr. Morgan, not sure why, but Judge Peterson felt it was important enough to send us after you. Could you please come along?" Then she took a brief formal mood, "Besides I would not like to have to use my stun gun this close to the water, I might get shocked."

Charlie laughed at the marshal's comments as he closed his back pack. While waiting for Charlie the two marshals looked into the deep clear water. The sun was now penetrating to the bottom.

The young lady pointed towards the creek and asked the other officer. "Down there…is that a golden trout? It doesn't seem to be moving. Do they hide like that on the bottom?"

The man leaned toward the creek to see if he could tell anything about the fish lying still. "Not sure what it's doing. I guess it could be just holding position waiting for a meal to float by, or it could be dead. Looks to be a big fish, may just be an old brooder that died. Hatcheries release the old ones into the streams in hopes some fisherman might catch themselves a trophy fish."

He turned to Charlie. "Mr. Morgan… you have any luck today? This far back in the canyon would be pretty good fishing I'm guessing."

Charlie thought for a moment as he looked at the two. He had a quick thought to himself, "If I tell them what I did, maybe he'll talk her into diving in to retrieve the evidence." He thought for a moment before he answered the marshal's question, "Not today, just junk fish biting today." Charlie took one last glance at the secret fishing hole and wondered if he would ever see it again. His knees made it increasingly difficult for him to walk in this far.

He turned and said to the young man, "Here, carry my back pack, you're in good shape, better than a 71 year old lawyer with bad knees."

The marshal picked up the back pack and swung it over one shoulder. The young woman took a hold of Charlie's arm to help him along. He really did not need her help, but he was not going to refuse a lady with a gun who offered to help him. He was both grateful and pleased. After a moment he realized these two had interrupted what may be the last time he went fishing for "The General." He turned back to where he had been fishing and thought. "Maybe God will allow me one more trip to this place. If my knees hold out a while longer."

The two marshals and Charlie started down the trail leading back to camp.

No more would be said of the Great Golden Trout dead at the bottom of Smoke Hole Canyon.

Chapter 3
Judgment by Peterson

The trip back to camp took over an hour even with the two marshals helping Charlie along the path. When they returned to camp, they assisted him gathering his things and packed them neatly into the back of his truck.

"I don't suppose you ever had an assignment like this before? By the way, what are your first names?"

"My name is Samantha; I mean U.S. Marshall Harris. I have been with the marshal service for three years, and no, I never had an assignment like this one. Do you mind me asking… are you someone important, sir?"

Charlie laughed, "No, I'm nobody now. Really never been anyone that you would have heard of. Spent most of my years working as a special prosecutor for federal crimes that involved corrupt officials or high ranking private businessmen. Somehow I started on that track when I was a young assistant prosecutor and life took me in that direction. Lived in Elkins most of my life. A few years after my wife was killed, I moved to a cabin we kept near Petersburg. I spend most of my time there, quiet and peaceful. It's just me now, don't make much difference where I go to bed. Truth is, I'm most comfortable in the solitude of this canyon. At least once a year I come back here for a week or two. Now that I am retired I could come more often if I wanted. Trouble is if I come too often, it might lose being special to me. So, I plan all year to make this one trip to catch one special trout. I just need a tent to keep me dry and the bugs off along with a good sleeping bag. I keep a few books in my truck to read and spend the evenings listening to the crickets and coyotes while I read Zane Gray and Jackie Collins by firelight. I may not have sex anymore, but what I remember I don't think I ever had sex as much as the characters in her books do. And Zane's books don't challenge my mind to understand the story's plot. I enjoy the good guys killing the bad guys and getting the girl in the end. I wish some of Zane's heroes had been around to take care of some of them worthless bastard I locked away. Some deserved to be shot at high noon in the dusty street of Elkins. At least their worthless lives would be good for something in one of his books. And like most of his good guys, I like it when the man in the white hat gets the girl in the end. Bad guy dead and hero grinning from ear to ear. Ah… they don't tellum' or writeum' like that anymore."

Charlie laughed as he waited for a response from the two marshals. "Well… don't you two agree?" he said. They pretended not to hear his comments.

Charlie grumbled for a moment and then said, "Other than reading, I enjoy catching my supper from the stream. I just…well I guess all I really do is try and enjoy every day I have left. Out here I bother no one… and I'm damn sure I don't want any one bothering me. Hell, I can't even do that on my own now. You two are evidence of that. I don't want to be special. I don't want anything but to be left alone. And I suspect you two are not going to let me do that, are you?"

Charlie waited for an answer. Before they could figure a nice way to tell him no, he asked them another question.

"And what about you?" He pointed his finger at the man.

"Well sir, my name is Jud Hickman, named after my grandfather. He was a game warden in this area when I was younger. Got lucky when I was assigned to this district. I cover this territory with four other marshals. Getting an assignment like coming to find you was a surprise. Normally they would have sent local cops, but the judge figured you might piss them off and they would shoot you. Besides, I guess they figured since I spent time with my grandfather when I was younger, I knew something about this canyon."

Jud looked down at Samantha and said, "I reckon they did not want a road cop or a woman getting lost in the wilderness." Samantha punched him in the arm.

Charlie laughed out loud, "A girl with spunk, I like that."

His comment turned her attention back to him as they loaded the last of his camping materials into the back of his truck. "Well, Mr. Morgan, just why do you rate two marshals trailing you into the wilderness, just to bring you out to visit your friend, Stinky… Judge Peterson. Don't tell him I said that!" She covered her mouth with her hand as if that would prevent her from saying anything else that might get her into trouble.

"Don't know any more than you do Sam, He may need a fourth for a hearts tournament at the club, just don't know. I can't imagine why he needs me either. We haven't spoken in a couple of years. Just no reason too since I retired. Well, I'll get my stuff together and head out when I'm ready. I'm sure you two have other important business, and I don't want to delay you any more than I already have."

Jud held out his hand, "Sir, I'll take the keys to your truck. Our orders are to bring you back. I think the judge said, "Don't let the grouchy old bastard out of your sight. He'll take off on you." I'll be driving your truck to Elkins, you'll be riding with Marshall Harris. You be a good fellow and she may let you turn on the lights and siren."

Charlie ignored the part about the lights and siren, "what do you mean you'll drive my truck back. I'm perfectly able to drive myself."

Samantha spoke, "we have our orders, bring you directly to the court house, don't let you head off home or somewhere else. We figure the best way to insure that happens is for Jud to drive your truck and I take you in our Suburban. Now, are you going to ride up front with me like a good retired federal prosecutor, or are you going to ride in the back seat like a grouchy old fisherman… what will it be?"

Charlie did not like it, but he knew the young marshals would do their duty and all fun would be set aside. Whoever sent them knew they would carry out their orders to the letter.

Charlie pursed his lips with a foul look as he tossed Jud the keys, Watch the damned clutch, it sticks sometimes … have to use your foot to pull it all the way up." He looked at Jud, "You can drive a standard, can't you?"

Jud never responded as he got in the truck, started it and headed off down the road. Charlie looked at Samantha, "You know Sam…I like him, a man of few words… kind of like me. I didn't see a ring on his finger…I reckon he's not married."

He looked at Samantha's hand, "Well, I see you have not got the ring of marriage either. I notice you two make a fine couple, any chance I can get you all hitched up, I know a federal judge who'll take care of that this afternoon?"

Samantha removed her jacket and placed it in the back seat, "Sir, with all due respect, shut up and get in … and by the way hook up your seat belt or I'll give you a ticket." She smiled as Charlie got into the black Suburban and they headed off down the dusty road that led out of the canyon.

The trip back to Elkins was going to be long. No direct route from the canyon and the only way back was also the quickest way. Head towards Franklin, get on 33 and head over the mountains toward Elkins.

Charlie and Samantha talked most of the way. Nothing important, mostly talk of family and the sights along the road. Charlie never said much about his past, he was mostly interested in Samantha and her life. He asked her how she became a marshal.

She explained, "When I got out of high school, there were no real jobs in this part of the state. So, I decided to go into the army reserves and attend college. I needed the extra money to pay for the schooling. I ended up in the military police and liked the job. Learned that the assignment was clear, enforce the law… that was the easy part. The part that took a while was knowing how and when to enforce the law."

Charlie sat for a moment and then looked at Samantha, "What do you mean, when to enforce the law?"

Sam thought for a moment before answering, "During the summer my unit was sent on maneuvers to a base down south. Our unit's assignment was to police traffic on base. I was a road cop for lack of better terms. I would turn on my radar unit to check the speed of traffic in the base housing area. It was a place where it was not uncommon for people to be walking and kids playing. If you were traveling 5 miles over the speed limit in a residential area, I gave you a ticket. I gave you no room to endanger some kid on a bicycle. But, the next day my duty assignment may have been normal traffic patrol on base. If my radar picked up cars in traffic going over the speed limit and the flow was normal, I cut them some slack. I learned from that job, a good officer needs to use common sense when enforcing the law. Then when I was leaving the army, I saw on the information board at headquarters, the US Marshall Service was looking for recruits with experience. I figured it might be a job I could work until I finished college. I wanted to be a lawyer if I could get into school. I figured by being a lawyer, I could make a difference in the world. I know …big ideals."

Samantha looked over at Charlie, "Are you a person of big ideals? Did you make a difference? Is that why the judge sent us to get you?"

Charlie sat for a few moments before answering. "I like to think I am, or was before I retired. I got old and tired of trying to find all the truth when prosecuting a case. Half-truths and political agenda were part of many big prosecution cases, especially when the government had an interest in the case. The government usually had their own agenda. For me it was all about, 'are they guilty or not?' And then after the government investigators told me their case, I had to deal with the criminals and see if what they said was the truth. In cases dealing with Wall Street, big government or major defense contractors, everybody has an agenda. I just got tired of the back stabbing and just wanted to catch fish. I use to love the law… but somewhere along the way, shysters got into the law. It was a game for them to see how much money they could make. Right or wrong got lost in what could be personally gained. A big law firm that represents a major defense contract has their own lobbyist. If they are representing a client with multi-billion dollar contracts, law firm's fees are staggering. You say billions when describing their money and people would sell out their own mother, wife and God. Justice is all about the money. Hell, when my investigators went to work putting together a case, the first thing they did was, follow the money." Charlie just sat and said nothing for a few minutes.

After a while he looked at Samantha, "Now, what about you and Jud, does he excite you? I see the way you two look at each other when you think I'm not lookin'."

Samantha laughed. "Leave it alone Charlie. We're just doing our jobs. Besides, before this morning I hadn't talked with him since college. Well, at least we have never talked outside of our duties"

Charlie sat for a moment before grunting, "Uh-huh. Whatever you say Sam."

After that neither said anything for a while, then Charlie said, "When my wife was alive, I like to think I did things to make it a better world. Spent long hours studying the law and preparing every case." He looked at Samantha, "I hate to lose at anything."

He turned back to looking at the highway, "But, when she and my son were killed I realized that as hard as I tried, I could not have prevented their deaths. The man that killed them had been arrested by the law many times. But, loop holes in the law and having enough money to find a lawyer who could find those holes in the law is what really killed my family. I decided to retire and forget making the law do its job. Even when I did the best I could, money or politics sometimes changed things. I fish now and read books. Neither can hurt me anymore. And now young lady, you and Jud pulled me out of my canyon where I was trying to catch a great rainbow trout. Most likely the last time I will make the trip into that secret place. Knees are going bad, doctor says I need both of my knees replaced and maybe a hip. I'm not sure I want them operated on. Even if I do, I'm not sure these old legs can carry me into that place again. I refuse to fish from a handicapped wheel chair ramp."

He turned to Samantha, "You know, last year I stopped at one of those places to see what the water looked like below the ramp. On the corner of the platform was a man in a wheel chair. He had a couple of small flags attached to the back of his chair. One was a black MIA POW flag. The other was an American flag. I knew he was an American Veteran. He sat there trying to cast his line to a spot where a trout may have been hiding across the creek behind a rock. As I looked down at the creek I saw three men and a couple of kids playing on inter tubes in the fishing hole. They were having a grand old time. They took no notice of the man trying to catch a fish. Those ass-holes had the whole stream to play in, but they chose that spot. The man in the wheelchair had nowhere else to fish. But, you know what Samantha, he kept casting; he hoped to catch a fish and the assholes below were not going to stop him. I had to admire any man with that much courage. I don't have that much. I'd be tossing M-80s down on them assholes swimming below."

Charlie laughed, "Be like depth charging the enemy, wouldn't it. I think I'd enjoy that." Samantha laughed and made no comment about his remarks.

Chapter 4

Elkins

It was nearly 5:30 when the three arrived in the town of Elkins. Charlie had been asleep for the last half hour. A couple of time Samantha looked over at him and he seemed to be hanging by his seat belt. She wondered if he were still alive. Then a bump in the road would awakened Charlie. He looked around before repositioning himself to be more comfortable. As they came down the long hill into the edge of town, Charlie spoke without opening his eyes. "Take me by my house, it's on Elm Drive, I want to take a shower before going by the court house."

Samantha thought before answering, "Sorry Sir, our orders are specific, find you and bring you to the Federal Courthouse. The judge said not to let you detour us away from the meeting with him. He will be there awaiting your arrival. Sorry, but orders are orders and the judge was very specific in his directions. His exact words, "Don't let him bamboozle you into avoiding the courthouse." Besides, in ten minutes we will be there, you can figure it out with the judge."

Charlie let out a deep sigh of air as he rolled his lower lip over his top. Clearly he was not happy. The nap had made him even grouchier.

Traffic was still heavy as they turned onto the busy street. They could now see the court house. The Randolph Federal Center. Charlie pointed toward the building. "Do you know who Randolph was?"

Samantha hesitated for a minute before answering. "I know he was a politician, beyond that, no."

Charlie said nothing as they pulled into the lot next to his truck that Jud was now parking.

As Jud stepped out of the truck, he stretched and twisted his body from the long ride. He then put on his suit coat and adjusted his tie.

As Charlie exited the SUV he asked Jud, "any problems with the clutch?"

Jud smiled, "No problems, I worked on the farm during the summer at my grandparents. I drove their tractor when they were making hay. Grandpa's tractor had a clutch. Never had a problem with his. Fact is yours never stuck on me, are you sure you don't ride the clutch and cause it to get hot and stick?"

Charlie looked at Jud, "Ride the clutch… ride the dam clutch. Let me tell you that I don't ride the clutch. I might rest my foot on it, but I don't ride it, give me my damned keys!"

Jud held them high overhead, "Uh'uh'uh. Not yet, we need to see you into the judge's chambers before I can give them to you. Orders you know." He chuckled.

Charlie mumbled something like, "what ever happened to law men being civil to private citizens? I KNOW… you all got guns, a poor old man like me is just being man handled by the two new lawmen of the land. Ain't that about right?"

He looked at Samantha as she was now exiting the Suburban. "Oh excuse me… LAW PERSONS. Is that the term politically correct these days?"

Samantha looked at Charlie, "You know, I could shoot you in the foot and Jud would swear you were trying to run…no one would question the word of two new age LAW PERSON's. Especially if I shot an old grouch."

Charlie grumbled some inaudible words as he started toward the back steps of the courthouse. As they went along Jud said to Charlie, "You better get your brakes checked. Coming down the mountain I think I could hear metal on metal when they got a little hot. When was the last time you checked them?"

Charlie never answered, he just hurried his pace toward a large metal door. A sign hung to the left of the secure entrance into the building. On the sign were instructions for those wanting to enter. "Please press buzzer, step back so you can be identified on security cameras. NO EXCEPTIONS."

Samantha pushed the button and the three looked up at the camera. In a moment a stern voice came over a small speaker. The sound of the voice was as if it came from a tin can.

"You in the dark hunters cap, remove your hat so proper identification can be made."

Charlie pulled off his hat in frustration, "THERE… can you see me now. I don't want to be here, but your boss upstairs had me drug here under extreme duress, now open the damned door!"

The speaker crackled for moment. "Well, well, is that you and your warm personality Mr. Morgan?"

Charlie put his hat back on and did not answer the tin can voice. A metallic buzzer sounded and snap of the lock was heard. Jud pulled open the heavy door and motioned for Charlie and Samantha to enter.

As they started down the well-lit hall way, Charlie saw a door open just ahead. He could hear the sounds of machinery coming from inside. He stopped and looked in. The sign outside the door said, HVAC equipment room. He looked around until he saw a man writing on a clip board. He called out to him. "HEY FRANK."

The man turned and smiled as he quickly came toward the three standing in the hall. As he approached, he indicated for the trio to step back from the open door. As he came out into the hallway he closed it behind him killing the high pitch sound of the equipment. He turned towards Charlie as he removed one of the yellow foam plugs from his ear.

"How the hell you doing? I ain't seen you for, what has it been three, four years? Figured you moved south or something."

Frank was the only man Charlie seemed to enjoy talking with, maybe because he gave it to Charlie as good as he got. They also like to talk about making trout flies to fish with. Frank had plans of making flies and fishing plugs when he retired. Make some money and do something he liked. Trouble was, his daughter got hooked on prescription drugs after falling during a gymnastics routine. The drugs at first helped with the pain. Somewhere along the line the drugs took over her life. She had been in and out of treatment centers for the last five years. Frank had to work to help with the expenses. He and his wife had sacrificed everything to help their daughter. It had not made much difference. She seemed to always return to the drugs after a short time.

Before Charlie retired he got their daughter into a special rehabilitation center. He pulled strings and got the cost offset by a pharmaceutical company he said caused the condition by not properly warning the family of the possible consequences of the drug's use. Frank's daughter came out with the habit controlled. Within six months she was right back where she started. She was helped with the substance abuse, but when she got out her friends still had the problem. Hard to escape when those around her were dealing and using.

Charlie grabbed Franks hand and shook it hard. He did not ask about the family he just said, "I almost had "The General" on my hook this morning… that was until these two officers pulled me out of the canyon and brought me back here."

Frank removed the second yellow foam ear plug from his other ear as he said. "Someday Charlie, I'll be goin' with you and we can catch that bastard. Well, at least we can sit around the camp fire and tell the stories of the ones that got away." They both laughed.

About then Frank heard a change in the pitch of the chillers compressor from inside the room. "Charlie, I gots' to go. That damned chiller is acting up and I gots' to have a talk with it. The only reason that piece of shit unit is still running is because I know where to kick it in the ass to keep it going. The building's projects manager planned on replacing it with a new unit a couple of years ago. No money in the budget for a big project like that. I've got to keep it workin', you know if I don't, it gets damned hot up in them court rooms with all the hot air you lawyers put out." Frank shook Charlie's hand again and opened the door into the equipment room. Charlie could hear the compressor surging as it struggled to run.

Samantha touched Charlie's arm, "Sir, we must go. Security will have told the judge we are in the building."

They moved down the hallway until they came to the elevator's doors. Jud pushed the call button. In a moment the doors opened into a shiny well-lit elevator. The doors closed in a moment. Soft music was playing inside the sterile looking elevator.

Charlie pointed toward the camera. "See that, there is a guy watching us that knows I hate Barry Manilow music. He starts it playing every time he is on duty… just my luck he has the duty today." Charlie looks up at the camera and gives the glass eye the finger.

Samantha nudges him, "That is not very nice. You don't know who's watching."

Charlie, still staring at the camera said, "I don't give a shit!" About then the doors opened into a large hallway.

Charlie wasted no time as he bolted out the doors towards the judge's chambers. Halfway down the hall, a security check point was manned by one guard. The check point was positioned to block the hallway. Passage was possible only through one of two metal detectors.

The guard looked up at the three approaching the checkpoint. "Can I help you all?"

Jud spoke first, "We are here to see Judge Peterson and he is expecting us."

The guard picked up his phone and pushed a button. In a moment someone came on the phone. "Yes, this is Officer Berger; I have two marshals escorting an older gentleman and they say they have an appointment with the judge. Can you confirm that, please?"

They could hear the voice on the other end say something, then the guard looked at Charlie. "Sir, are you Charles Morgan?"

Charlie leaned forward and said, "Tell the judge the longer I stand here the meaner I get. I should be fishing and not listening to Barry Manilow music. Is that enough identification for you Mr. Berger?"

The guard's attitude changed as he stood holding the phone to his ear. "Yes sir, you all may go on in, the judge is expecting you in his chambers. I am pleased to meet you Mr. Morgan. Your reputation proceeds you, and oh yes, I know how you love a good Barry Manilow song. I've saved them just for you." The guard smiled coyly as Charlie once again grumbled inaudibly.

The three stepped through the metal detectors. The two officers set off the device and the guard motioned them through. When Charlie stepped through, the alarm sounded again. Clearly the guard was a bit perplexed.

"Sir, do you have any weapons or metal that may pose a danger?"

Charlie looked at the guard as he reached into his pocket and pulled out a small pocket knife and a metal fire starter stick. "Here, I got a knife, you do know most real men in West Virginia carry some kind of pocket knife don't you, Mr. Berger?"

The guard pointed for him to pass through the detector one more time. As he did it sounded once again.

Everyone was now becoming more inpatient with the security alarms. Samantha stepped toward Charlie, "Quit fooling around, what else have you got on you?"

Charlie smiled and reached inside his coat pocket and removed a metal flask. "The officer just asked if I had anything that posed a danger to others." Samantha just shook her head and grabbed him by the arm as he laid it on the table.

As he started to step through the detector one more time he stopped and looked back at the guard and said, "I know just how much Jack Daniels is left in that flask, you drink any and I'll know!"

The guard was unimpressed with Charlie's threat. Finally he passed through the metal detector and no alarms were sounded. They moved along the hallway until they came to the door with Judge Peterson's name beside it. Jud opened the door and the other two went inside. Sitting at the desk in the outer office was a woman with bright red hair busy typing as she listened to headphones. She was transcribing notes the judge most likely needed the next day. The door to the judge's office was opened and the three could see the judge sitting at his desk.

In a booming voice the judge called out to Charlie, "Well you old grouch! Thought you could hide from me in the mountains. I know you better than that. And I knew you would not come if I just asked politely… So." He pointed to the marshals. "I sent the cavalry, knew you would come that way."

He came around the desk and stood near the door, "Marshalls, I thank you, I'll make sure it is noted that you two performed above and beyond the call of duty to get this man here. Now, if you will excuse us, we have important business to discuss."

The two started for the door. They knew the judge wanted them to go and that was his way of saying good-bye without explanation for their assignment. As Samantha neared the door she turned and stepped back toward Charlie. She gave him a questioning look, "Who… are… you? I can tell by the people you meet you are…or were somebody worth knowing. That is after you get by the crustiness on the outside. And now it looks as if I'm not going to have that chance. Good luck Mr. Morgan and maybe someday I'll give you a ticket." She smiled. "Just a way to say howdy."

Charlie shook her hand and smiled as he looked over at Jud. "Son, you two already have your own handcuffs, you ought to ask this lady to play cops and robbers." Samantha slapped him on the arm in response to his comment.

"That's enough of that kind of talk," she said. The two marshals stepped out of the judge's office as Charlie watched them go before Peterson closed the door.

"You know, those two are good kids. I think they like each other, they just ain't figured it out yet."

Charlie turned toward Peterson. "Well, what the hell is going on? In all my years of practicing the law, this has never happen to me before. You better have a good explanation for having those marshals kidnap me and drag me back here. I'm retired. You have no jurisdiction over me. Bill, what's this all about?"

Peterson turned the lock on the door and headed to a cabinet near the front wall. He opened the door and inside was a bottle of Jack Daniels along with a couple of glasses.

"Charlie, do you remember when we were first starting out? We were two young lawyers with big ideas about the law and how we were going to bring justice to the courts. We figured with our new suits from Sears and Roebuck's, and arms full of new law books, we could be the next Clarence Darrow? You had a real love for going after the truth and finding corruption. You were good, damned good. The mere mention of your name as the prosecutor in cases, often sent the defense to ask for a deal." Peterson pointed his finger at Charlie as he poured him a drink with his other hand.

"And you…you were like a pit bull. They may have gotten a deal, but they lost a leg or arm in the deal. Yea, Charlie you were good. You had a real love for it. You loved going into court and using the law to find ways to put criminals away. Especially white collared crimes. The smarter they thought they were, the more you enjoyed pulling them down."

Peterson handed Charlie a glass as they both sat down. "Bill that was a long time ago. Water under the bridge. I loved the law like you said. In fact, so much so I forgot about my family. I spent time with them, but I was usually thinking about a case. The only time I really ever forgot about the job was when I was fishing in the canyon. The high walls kept out the justice system." Charlie took a small drink as he looked at Bill.

"Bill, I never asked what happened to you and that pit bull attitude that you came here with?"

"I didn't like the tediousness of the everyday world of prosecution. The details of the evidence that had to be verified. When I got the opportunity to teach back at the university, I jumped at it. I thought it would fill the need I had for the law. Tried that for a few years. One day after class, two men came into my office and asked me to join them for dinner, they had a proposal they wanted me to think over. They were from the Justice Department. The judge for the Northern District was going to retire with sudden health issues, bad heart I think. The U.S. Attorney General appointed me to fill the position temporarily. I saw it as something new, maybe a judgeship would be something I could enjoy. I do like the law. As they say the rest is history."

Charlie took another drink from his glass. He chuckled, "Did you feel you would like being a judge before or after the mining case?"

Peterson looked at Charlie with a stern look, "You could not let that go could you, that was a long time ago and you just can't forget it."

Charlie took another drink and stood up as he walked to the window. "Now, that I got you pissed, what the hell do you want, Bill? You did not bring me here just to have me stir shit, not after all this time! There is something you really want and you want it bad enough to ask me. Let's face it we were friends when we were young, but I did not agree with some of your judgments from the bench. And I know you did not like my pig headedness when it came to prosecuting cases in your court. I saw the law as being black and white. You and the courts say it could be shades of gray when it came to corporate America. I didn't see it that way. The little guy steals $200 from a gas station to feed his kids. He gets 20 years. Sure, that's what the law says he should get. But, when I prosecuted a Wall Street guy who cheated millions from pension funds, you gave him five years in a minimum security prison. What kind of a suit the accused walked into court wearing made a difference in your sentencing. You forgot we walked into court wearing off the hanger suits from Sears. No, I was not a fan of your sentencing guidelines. That's why I worked so hard to make sure they were guilty with no room for your soft sentencing terms."

Peterson stood up, "That's bullshit Charlie and you know it. You know how the law works. It was easy for you, guilty or not. Well, for me that was never the way it was. If you all in the prosecutor's office saw a way to let a little fish off the hook to get a bigger prize, you had no trouble cutting a deal and coming to me to sign off on it… ain't that right?"

Peterson pointed his finger at Charlie, "You remember that case where the drug pusher kidnapped that young girl to cover his getaway. That young girl was the daughter of one of the east coast's biggest drug suppliers. You guys in the prosecutor's office offered him a deal to give up the supply's network in exchange for witness protection. You remember the little girl Charlie, don't you? Well I do… I still remember the pictures of her burned body. What was it your people said, there was no way to prove the suspect killed her and then burned the body. You all were in such a big hurry to convict the supplier, everything else was forgotten, including that little girl. No Charlie, we all have skeletons in our closet."

Peterson sat down in his chair and said nothing for a few moments. Charlie stood looking out the window.

Finally Peterson said, "Charlie the law is only as good as the people who try and administer it. I wish it was black and white. I wish I could have sometimes asked you lawyers for more information before sentencing people. I think I wanted to be the judge who was administering the law without prejudice. But hell, I am only human, just like you. Do you know how many times I sat on the bench deciding what to do when a jury decided the case solely on evidence I allowed in or kept out? Evidence that may have been introduced by some legal maneuver you lawyers came up with. You know damned well I sat on the bench looking at the three people who probably knew the real truth in a case. The defense, his client and most likely you. Then I would look at the jury and knew the evidence I allowed in would make a difference in how they judged it. I did my best… still am. Tomorrow I'll hear another case. A case where evidence may only tell part of the truth. I do love the law, but I also understand the law is not always perfect."

Neither man said anything for a few moments. Finally Charlie said, "Did not mean to bring this all up Bill. We were once good friends. I am guessing you called me here because we once were and now you need my help. But, why the officers… and why send them after me out in the wilderness?"

Peterson opened the middle drawer to his desk and removed a framed picture. He leaned up in his chair and handed it to Charlie. In the picture was a much younger Peterson along with an attractive dark haired woman. The woman wore a white lab coat like a doctor would wear. Around her neck she wore a stethoscope. Charlie looked at the picture for a minute and handed it back to Bill.

"Who's the woman in the picture beside you, I know that's not your wife?"

Peterson studied the picture for a moment. "You remember when I returned to the university to teach, I met someone, a young doctor. Her name was Elizabeth Mathenson. She had dreams of someday changing the world of medicine. Every moment of the day was to be in pursuit of her dream. She had new and radical ideas how to change science and medicine by combining them. Elizabeth was studying microbiology and genetics, believing the two could work together to find cures for many diseases. She had some crazy ideas that all she needed was to find the key to unlock molecules in a person's genes, reprograming them to repair or replace defective genes. If she could do that, she believed her science could save lives. Most of the time I didn't have a clue as to what she was talking about, but she did."

Peterson returned the picture to his desk drawer. We had a great time together in the summer of 81, but she was dedicated to her research and I was… well I was teaching and hoping to find what I wanted to do with my life until the judgeship was offered to me. One evening we were discussing her research and the possibilities she hoped someday it would open. We disagreed on the outcome she hoped to bring with her science. My faith gives me unquestionable direction in my beliefs', I came to realize Elizabeth was not a person of faith, at least in

God. Shortly after that we began drifting apart. Then we decided to both go our own ways. No hard feelings, we parted as friends. I never heard anything from her for several years after I left the campus, then I was asked to return to the university to speak to a law class. While I was there, we, by chance, met once again. She was still as driven as ever to find answers. Only now she had a full position in the science department and was heading up her own research project in bio-genetics. I later found out she was considered one of the very top scientists in her field. We talked and laughed about old times. I never really questioned the direction of her reseach. I told her I was married and had two wonderful kids. That seemed to touch something within her. I could see it in her eyes. But, she smiled and we talked for a while longer until I had to return home. Never saw or heard from her again. Until I learned warrants and subpoenas had been issued by the U.S. Attorney's office. They alleged, Elizabeth misappropriated and embezzled government funds and withheld material evidence. The subpoenas and warrants are very broad. Looks to me like they are on a fishing trip. In fact, they are looking for something specific using a very broad legal net. The warrant is looking for the normal evidence you would expect, but it is also looking for biologic materials. Something called a cryogenic chamber. Other than that, it gives very little information about the case. Asked around if she has a lawyer. Found out the university is distancing itself from her and she has a young lawyer just out of college. I can't find much else out without raising legal questions about my interest. That's where you come in, you're retired and you know the legal system when it comes to embezzlement cases. You and your field investigators could find a penny hidden under the rug in these kind of cases. I brought you here wanting to know if you could check into this for me without raising any suspicion. Face it, most people are not going to put you and me working together on anything."

Charlie removed his hat and scratched his head. He looks up at Peterson. "Jesus Christ Bill, this is what you pulled me in here for. A damned fraud case with some old girlfriend?"

He stood up and walked over and filled his glass from the whiskey bottle and then poured Peterson's glass full.

Best you get the prosecutor on the case to think about a deal. Tell him… tell him… hell, I don't know what to tell him. Bill this case is none of your business and if you get involved it could be a world of shit for you and the woman, what was her name?"

"Elizabeth. She was someone special to me at a time in my life when I was lost as to what I wanted to do. She came into my life and helped me see the way I should go. I owe her for that. And damn it, you can help me. You are the best when it comes to prosecuting a case like this. You understand the law. If you were to be her attorney it would give her a fighting chance to come through this without going to jail."

Charlie turned quickly and looked at Peterson, "Is she guilty? The first thing I have to know, is she guilty?"

Peterson paused for a moment, "I don't know."

Charlie chuckled out loud, "Well isn't that just great! You want me to defend a woman and you don't even know if she is guilty. Have you even talked with her?"

"NO, ABSOLUTLY NOT! I did not want to take that chance until I talked with you. And even then, because it will be a federal case and I'm a federal judge, I have to be careful. You know them damned federal prosecutors, they'll try and drag me into this if they get a chance. Some of them fellows would love to try and make me an accessory after the fact. Be a big feather in their career caps if they can implicate a federal judge. I promise you this, I am clean in this matter and I just want an old friend to get a fair shake when she goes to trial."

Charlie turned and looked at Bill, "So you're telling me, you have absolutely no involvement with this woman since you left the university?" He steps closer to the judge, "Have you talked with her at all since this came up? Contact of any kind. I don't care if you just smiled at her across a room full of people, I have to know you are not involved in any way?"

Peterson never blinked as he stared at Charlie, "I haven't spoken to her since that time years ago when I went back to speak to the law class. Like I said, if I had not seen her name on the indictment docket I would not even know she was still alive."

Charlie sat down and rubbed his chin. "Why am I even thinking about this? I'm retired. I am happy. Well, maybe not totally happy but I don't have to worry about this kind of bull shit anymore."

He stood and started toward the door. As he went he said, "No… no, I am not going to get myself involved in any more legal cases. Besides, it is just a simple embezzlement case of federal funds. It won't draw any attention. If she is convicted she'll get five to ten years. Four-and-a-half with good behavior. No Bill, I can't do it. Get someone else."

Charlie reached for the door knob when Bill said, "Charlie she has a son, his name is Robert. He is 35 now. He is my son."

Charlie froze as he held the door knob and without looking at Bill he asked. "Does your wife know?"

"Yes, she knows. She knows he was born before we even met. Beth was a college fling when she got pregnant. She never told me until 10 years later when I returned to speak at the university. She never asked me for anything and, believe me, I offered. She's an independent woman and raised him on her own. There is nothing hidden here in my family or hers. She told me she was going to tell her son when he was older. This request for you to help her is me just wanting to help the lady from a time many years ago in our lives. And yes, she just happens to be the mother of my son."

Charlie turned to Bill, "How does her son feel about all this?"

Bill sat down at his desk, "I don't know. When we spoke all those years ago he was young and she had not decided when she wanted to tell him. Not sure to protect him or to protect my reputation. She knew by then I was married and filling in on the federal bench. My appointment had not yet become permanent. A story of an illegitimate son could affect my family and my appointment to the bench. I am guessing she waited until he was an adult. Maybe he would be more understanding of that time in both our lives. But, I don't know that for sure."

"Shit Bill, you are making this hard. You got any more surprises for me?"

"No, I don't think so."

The judge walked close to Charlie and laid his hand on his shoulder. "Please, will you at least talk to her before you decide? And if you do, please don't tell her I had anything to do with you getting involved?"

Charlie thought for a moment. "If I do this, and I mean "IF", I can't promise not to tell her how I got involved. And I will need to bring Jake out of retirement too. But, don't send those two marshals to get him, he is paranoid enough the way it is. Two feds show up at his place will just reinforce his fears of the government spying on him. I'll go see him. I think he is still living up near Deep Creek in his lake cabin. Heard he was helping senior citizens improve computer skills at some seniors home."

The judge looked at Charlie, "You mean he is helping them learn how to hack computer systems?"

"I think Jake called it, deep investigation of computer viruses. At least that's what he told people when he got caught, which was only a few times."

Charlie and the judge said nothing for a few minutes. Then Charlie said, "I'll tell you what Bill, I'll talk with the woman and then her lawyer. I don't know what I'll say as to how I came to be interested in this case…I guess I could tell them I am tired of retirement and want to get back into the court room. While I am doing that, I'll have Jake try and find out what information is out there about your doctor and the money she is accused of embezzling."

Charlie pointed his finger at the Judge in a stern manner, "I'll tell you this, if I find any hint that she knew about or was part of embezzling money with criminal intent, I'll walk away. I'll walk so far from this, those marshals you sent after me won't find me. Is that clear?"

Peterson shook his head he understood.

"Charlie, I had no one else I could ask. I know maybe I should not ask at all, but at one time in my life she meant a lot to me. And she is the mother of my son. I guess in some ways I hope that you can make things right in their lives. Do you need anything from me?"

"No, absolutely not. From here on out, until this is over you are out of it, do you understand? So far you have not prejudiced yourself or me. I want it to stay that way. I am going to find out if this is a simple embezzlement case and hope that it draws no attention from the newspapers. You know reporters, if they get a sniff she is an old girlfriend of a now sitting federal judge. Put you two together and discover the age of her son. The case goes from page 10, to the headline of that day. No, you stay away from this. In fact, you don't look good to me, have you been feeling well lately? I think you need to take some time off, rest up, play a little golf. If you're under the weather, they'll have to bring a judge from another district. Or send it to one of the other courts. My guess is, they will assign the case to Wheeling or Martinsburg. No questions will be asked if you were needing time off from your busy schedule. But, if you recuse yourself from this case, the question will be, why? I think you should do it in the next couple of days. I am sure you have sick leave built up. Take your wife and go to the Greenbrier for a while, at least until the docket date is set in another court. Recuse yourself and they'll be questions… it will also complicate it for your doctor friend and for you."

The judge said, "But Charlie, I have not done anything that could influence the case. If I simply remove myself from the case that will be the end of it, I have done that before when I had cause not to preside over a case."

Charlie looked at him as he said, "Yea, but you didn't have sex or a kid with those defendants did you? No, you take time off and spend it at the Greenbrier or, even better, go to Myrtle Beach, they have some great golf courses there."

Bill shook his head that he would follow Charlie's suggestion.

The two men shook hands and Charlie started to leave. He stopped and turned to the judge.

"Won't people wonder about you and me meeting? Especially since you sent the troops to get me?"

"No. And as to those two marshals, I swore them into the marshal's service. Good kids. And the hard time you gave them, most likely they'll understand sometimes you are a pain in the ass. The other people here today, they'll not ask questions. Besides, if they do, I'll tell them we're old friends and go way back."

Charlie grinned and said, "Well, let's hope we're still friends when this is over. Remember, low profile. You get yourself some quiet time and stay out of this case. Myrtle Beach is nice this time of year, kids back in school and just old people vacationing before the winter weather sets in."

Chapter 5
Breakfast at Bob's

As Charlie left the judge's office he looked at the red haired lady with head phones on, listening to testimony and typing the words. She still gave no indication that she had noticed Charlie's arrival or his departure.

As he headed down the hallway toward the check point, he noticed the same guard awaiting him. Charlie stopped and gave him a stern look. "Where's my stuff, you didn't drink any of the whiskey… did you?"

The guard pointed for Charlie to pass through the metal detector. As he did the alarm went off, to Charlie's surprise. The guard grinned. "Sorry, I must have touched the test button."

The man slid a small plastic tray toward Charlie, who was now on the far side of the check point. "Sir… please check your items."

Charlie grumbled for just a moment as he picked up his pocket knife and fire starter stick. Then he picked up the flask. He held it in his hand as if he was weighing it. "It feels a little light, but I'll mark that up to evaporation."

The guard smiled as he said, "you have a nice day Mr. Morgan, oh, by the way, your ride down in the elevator…it will be made more pleasant by the wonderful sounds of Mr. Barry Manilow. You have a really nice day sir."

Charlie never said another word as he pushed the call button on the elevator. Almost instantly the tone sounded and the door opened. From inside came the sound of Barry singing, "Can't Smile Without You." He looked back at the guard who waved and smiled a big toothy grin. Charlie thought about giving him the finger, but he realized it was only music. Not his kind of music, but music. Besides, Charlie knew he would someday pay back Mr. Berger for the music. He might just tell a group of old ladies to be careful of the guard on the second floor. Rumors had it, he liked to get a little friendly when frisking them. He was sure the guard would appreciate his helpful advice to the women.

In a few moments Charlie arrived in the basement. The doors opened to a cool hallway as he headed toward the back door. As he passed the equipment room he could hear the muffled sounds from the equipment his friend was working on. Charlie hoped the man and his family could find help for their daughter.

In a moment he was at the door that allowed him to exit the building. Charlie opened the door into the bright sun causing him to flinch and cover his eyes. In the last few years his eyes were giving him more problems, especially in bright light and trying to drive at night. Charlie wondered if it was his age or maybe something wrong with his eyes. He had planned on going to an eye doctor, just had not gotten around to it.

As his eyes adjusted, he looked toward his truck. To his surprise he saw the two marshals waiting there.

"Did your meeting with the judge go well?" Jud asked.

Charlie fumbled for his keys as he said, "Well enough. What are you two still doing here? I figured you'd be long gone, back to protecting the citizens of this country?"

Samantha stepped toward Charlie, "Truth is we are curious, who are you? We know the judge, he swore both of us into the US Marshal's service. But you, you're a curiosity to me and I am a snoopy person. Must be my law enforcement background."

Charlie opened the door to his truck and as he closed it, he looked out at the two standing there with questioning looks on their faces. "You know young lady, curiosity killed the cat."

"Is that a threat Charlie Morgan?" she asked with a smile.

He returned the smile as he said, "No, Lord no, just not sure what you're wanting. The judge and I are just old friends, he had a pressing legal question about an old case that I once prosecuted. Nothing for you two to worry about. If I remember correctly, you two are stationed here in Elkins, that right?"

"That's right" Samantha responded.

"Well, before I go home I am going to get something to eat, you two interested in a free meal?"

They answered together, "Always."

Charlie said, "Follow me."

A short time later Charlie turned into "Bob's" restaurant parking lot followed closely by Jud and Samantha. The three went inside and were soon seated. Once the waitress had taken their orders, Charlie asked. "Okay, what are you two up to? You should have been long gone when I came out of the courthouse."

Jud took a drink of his coffee and as he sat the cup on the table he looked at Charlie and said, "You don't remember us, do you?"

Charlie looked at him with a questioning eye, "No… should I?"

"Probably not, it was a long time ago, about fifteen years. Our civics class was allowed into the courtroom during one of the cases you were prosecuting. Some company was trying to get rights to timber land in the National Forrest in a protected area. Not sure what all was going on, but we, Sam and me, were fascinated by your courtroom skills. We both planned on majoring in law at college after that day. After we gained our degrees, we applied to law school. They only accept so many applicants, neither of us made the cut into law school."

Samantha said, "Yea… you made a big impression on the both of us. You already know how I came to be in the marshal's service. As to Jud's story, he had completed pre-law and didn't get into law school either. He then applied for a game warden position. Shortly after that he was attending the state police academy. Then two years later, budget cuts put two dozen state police and wardens out of their jobs. Jud was one of them to be let go. A short time after that, he learned the marshal service was looking for new applicants with law enforcement experience…and, well the rest is history. Then we got this assignment out of the clear blue to go after you in the canyon. The attorney who sparked our interest in the law. This can't be chance or coincidence. Must be some bigger reason for our meeting after all those years."

Charlie took a slow drink of coffee from his cup. As he sat it down on the table he looked up at Jud and Samantha. "Fate… nah, it ain't nothing like that. It just is. People want to put too much into the stars aligning." Charlie waved his hands over his head as if he was some mystic astronomer.

"You're a couple of nice kids. And I'm just an old retired lawyer with a bad disposition toward people and the world. Seen too much and had some bad things happen in life. Maybe I should believe things happen for a reason. If I did that, I would wonder why I lost my family to an asshole with multiple DUI's. Nah, life just is and each day you do the best you can. No big secrets, no magic why man is here, he just is. And there is no God, cause if there were, my family would still be here. Now, enough about me." He motioned for the waitress, "Where's our food, I'm starving."

Jud and Samantha sat for a moment being blindsided by Charlie's revelations about his life and God. They sat and waited for the meals to be served, making small talk.

To break the tension, Samantha asked, "What's next for a retired federal prosecutor?"

"I have to go see a man in Maryland. He lives by Deep Creek Lake. Haven't seen him since he retired. I hope he will see me. With each passing year in his job, he became better at finding ways into the dark side of people's financial lives. But, at the same time, he began becoming more suspicious that others were doing the same thing to him. He believed they were watching him through his computer screen and listening to his phone calls. He became more and more paranoid as he began to believe the shadowy part of the cyber world was coming after him."

Jud askes, "Did you work together?"

"Kind of, he was an accountant who is a computer genius. He could find ways into computers and find hidden money. His official title was, Forensic Accountant. He was good, very good. Without his help, a lot of cases would have gone unsolved. To him it was like a game of hide and seek. He once told me, never put anything in a computer that you don't want someone to see. If you do, someone like me can find things you thought you erased a long time ago. Computers just pile stuff on top of stuff, it never really goes away if you know where to look. The deeper he got into the criminal's financial world, the better he liked it. About the same time I retired from prosecuting, he decided to spend more time on the lake with his wife, Ella. She died of breast cancer about a year after he retired. I did not hear about it until a few months later. I felt bad I had missed her funeral. I tried to call, never got an answer. Heard he was still living by the lake. Also heard he works at a senior center helping old people with their taxes. I've been thinking about paying him a visit. I will do that tomorrow. Now, if you two will excuse me, I'll pay for our meals and head home. I hope you won't be following me anymore."

Charlie stood and stretched his back muscles. As he started toward the register he stopped and looked back at the two still sitting there. "I know a guy over at the university in the law department. Maybe, and I do mean maybe… I could see if I could get you two into law school next term. If I could, are you interested?"

Jud and Samantha looked at each other and then back at Charlie. Samantha spoke first, "Yes, I know I would still like to become a lawyer, but I can't afford to go to school and be a marshal. That's one of the reasons I didn't apply a second time"

Jud hesitated for a moment before answering, "Charlie, why are you doing this, you only met the two of us a short time ago. Your offer is great, but our finances makes doing both very difficult at the present time, not to speak of the time needed for both. I don't think the marshal service wants part time marshals."

Charlie was fingering through his wallet as he said. "I'm not saying I can make it happen, just saying I'll check into it if you're interested. The Justice Department has been known to help people earn their law degrees, especially when they already work for the department. Of course they expect you to stay with the Justice department when you get your degree, at least for a while. They figure you are a known commodity to the department. And if you get your law degrees, they always need U.S. Attorneys. Most of the attorneys they hire today leave after a while to go into private practice. They have on their resume, Justice Department experience. That makes them an asset to firms that do business with the government. If you're interested, I'll see what I can do." Both Jud and Samantha smiled and shook their heads yes.

Chapter 6

Computers at the Bottom of the Lake

Charlie awoke several times during the night. His mind was already working the case and he was not even sure he wanted to be part of it. He did realize that when he was thinking about the case it was if he was the prosecutor. After a while he came to realize that perhaps if he took the case that could help him figure a better defense. He knew one of the reasons the judge had asked him, besides their old friendship, was the fact he was once a prosecutor. Maybe that would give him an edge. He finally convinced himself to clear his mind and fall asleep. His last thoughts, "You don't even know what the case is yet, and if the doctor even wanted his help. What if she asked him why I came out of the woodwork and offered my help?" Before he could rebut his own thoughts, he was taken by sleep.

The next morning the sun peeked through the blinds into his room. As he moved he felt the pains of sleeping on the ground for the last week. Somehow his bed brought out all those aches and pains.

He looked at the clock, quarter till eight. He wanted to be on the road by eight. It was not going to happen. He hoped to be at Deep Creek by ten or ten thirty at the latest.

After a shower and bowl of instant oatmeal, Charlie made his way to the truck sitting in the driveway. He noticed his grass had not been cut for a while. He paid the kid across the street good money to keep it cut. A neat mowed yard kept the neighbors off his back. They were always complaining about something. He lived in the housing development long before any of his present neighbors moved in. His old neighbors did not get all bent out of shape about the neighborhood agreement, which stated, *Lawns shall be neatly kept at all times.*

Charlie said out loud to no one in particular, "Screw-um, if they say anything about my grass, I'll sue them and fire the boy across the street."

Before long he was heading north toward the state line. It had been awhile since he had seen his friend. He really was not sure what he may be into these days, besides helping out at the senior center. Charlie mumbled, "That surely could not be too exciting. The strong smell of liniment probably could do in most men." He laughed at his own words.

It was nearly eleven when he saw the sign he was looking for, Sycamore Drive. It was a secluded lane that ran along the south side of the lake. Jake and Ella had inherited the property from her grandfather, who bought the property before the land became outrageous to own. The lot of land covered five acres. In today's world it was worth a small fortune on the lake front. Jake did not care about the money, he liked the seclusion.

After a few minutes searching, he saw a lane marker nearly hidden by the trees. It read, Sycamore Lane. Just below on the same post was a bigger sign that said, "**STAY OUT**." Charlie remembered his old friend had one problem that sometimes got in the way of doing his job, he was paranoid of almost everything, especially when it came to the government. He felt they could spy on anyone and you would never know. He believed his ability to electronically look into the government's world made him a big target. That is one of the reasons he and his wife lived off the grid along the lake.

As Charlie pulled into the driveway, he saw Jake standing near the end of his dock about ten feet out in the lake. He was throwing something into the water, it looked to be an old computer case. The man turned at the sound of Charlie's truck coming into the driveway. He strained to see who had trespassed onto his property and caught him throwing the computer into the lake.

"Who the hell are you? Didn't you see the damned sign to stay off the property? You from the damned government again?"

Charlie stepped out of the truck and held up his hands. Jake recognized his smile and held his hand up in a welcoming response.

"What the hell are you doing up here, Charlie? I thought you had gone on a quest to catch the biggest trout in West Virginia?"

Charlie smiled as he offered his hand to the man. "Good to see you Jake, What's it been, five years now?"

"Yea, about that. Come on up on the porch, I got a fresh pot of coffee brewing. You want one?"

Charlie shook his head that he did. In a couple of minutes Jake returned carrying the coffee. "If you want cream or sugar I'll get it for you." Charlie indicated to him he was fine with it being black.

"Before I forget, what the hell were you doing with the computer case when I drove up?"

"Now, you being a lawyer and all, you understand terminology is important. That was not computer trash… that was my effort to recycle those parts into a crappie fishing hole just off my dock. Those re-cycled cases make homes for little fish. Then the bigger fish come around to eat the little ones. Then I come along and catch those larger fish and I eat them. You see, I am helping nature and providing food for my table. I am part of the food chain. The top part"

Charlie laughed, "You missed your calling Jake, with words like that you could have been a good defense lawyer."

Jake opened his eyes wide as he said, "Watch your mouth Charlie, you could have at least said a good prosecutor and been nice about it."

The two men talked for a while, telling old courtroom stories. After a while Charlie even asked about Ella. Jake also mentioned how much he missed seeing Charlie's wife and his son. The mood between the two men became somber for the moment.

After about an hour of long missed conversations, Jake finally asked Charlie, "We've had a good talk, something I have not had in a good while. But, you didn't drive a couple hours to reminisce over old times with an old paranoid computer programmer or find out my secrets for catching crappie, what's up?"

Charlie thought for a moment. "Jake, you remember Judge Peterson from the Northern District?" Jake shook his head that he did.

"Well, he dragged me from my place in the canyon back to Elkins to check on a case for him. Nothing official and that goes for our meeting too. He wants me to check into a case that's coming before the court this session."

Jake looked at Charlie, "There is no conflict of interest here… is there? I don't want to get into some federal case that a judge has got his hands into and I get myself a load of problems. I don't need that anymore. I like being off the grid of the outside world. Getting involved in something like this could bring the government's attention to my five acres of paradise."

Charlie shook his head, "No, I don't think so. That's why he asked me to defend the person who is accused. He is afraid that if it is revealed he has some knowledge of the case, and the woman involved, he will taint the case."

"A woman, wait a minute Charlie. He ain't got woman problems he is trying to cover up?"

"No… well at least I don't think so. She is someone from his past when he was teaching at the university. Now the doctor's got herself into some trouble with misappropriations of grant money. Don't know the facts yet. That's why I came to you.

You're the best at finding the truth when it comes to money and people using federal funds to do the wrong thing."

Jake sat back in his chair as he looked out over the lake. "Charlie, I need a way into the system. A lot of security software has been put in place in the last few years. And if I understand what you're telling me, you have not even talked with this woman or her lawyer. We have no legal way to see the records of anything to do with the case. And you want me to find out if the money is there, or if it is gone, CORRECT? Charlie, if I bump into any fire walls trying to enter their system, they'll be on to me in a split second entering without authorization"

Charlie smiled for a moment before saying, "That's why I came to you, I am sure you have ways to move around their systems and not get caught. If I remember correctly, the CIA wanted you to come work for them doing just this kind of work. Besides, we're not going to take anything or change what we find. I just want to know what I'm getting myself into."

After a few more moments Charlie said, "I will give you the woman's name, its Elizabeth Mathenson. She's some sort of research doctor at Tygart University. Does most of her work with genetics. Anyway, she has done this for years. Now there is a question about the legality of her research or at least how she spent federal grant monies. For some reason the university has cut her loose with no tenure considerations. They have to know without extreme reason this will not be upheld with a legal challenge by the doctor. In fact it reeks of stupidity. That tells me the university reacted without thinking about the consequences. Or they know more than the public. That tells me there is more to the story than is known to the world. They would rather face criticism from the public over cutting her loose, than stand with her in whatever she has done."

Charlie paused for a moment. "And... as to the doctor, she hired some young lawyer fresh out of law school to defend her. For a supposedly intelligent person, she has done an extremely dumb thing to herself. Both the university and doctor seem to be going headlong into this controversy with no regards for a defendable position when it comes to trial. This shitty mess is what the judge wants me to help with and get in the middle of."

Jake chuckled as he said, "And you mean you're going to just walk into her lawyer's office and say. *Hey, I got nothing else to do kid, you need any help?* Bullshit Charlie. This doesn't sound like a good plan even to me. If he is any kind of attorney, new or old, he is going to be leery of an offer like that. Especially out of the clear blue. I may be paranoid about a lot of things around me, but you have to be a little paranoid too?"

Charlie looked at him and said, "No it doesn't. And on top of that I have never put together a defense for anyone. It's always been my job to tear the defense apart."

Charlie stood up and walked to the edge of the porch and stood quietly for a minute. From behind him Jake said, "I take it, if we do this it is Pro-Bono?"

Charlie laughed, what do you need money for, you got the best crappie hole on the lake, that's worth its weight in fish." He laughed.

Jake sat for a minute as he rolled his coffee cup around in his hands. Then he asked, "You say she has been doing this for a long time?"

Charlie shook his head yes, "The judge says she's been doing research for years on genetics. I figure she's been getting grants from the government and pharmaceutical companies. Oh yea, one more thing, the university or someone important locked up the research building and no one can even get close to see inside after her work was suspended"

Jake sat his cup down and disappeared into the house. In a few minutes he returned with a laptop computer. Charlie had never seen a laptop like this one. It looked like something high end gamers use to play on the internet.

Jake booted the computer and instantly it was up and running. "It makes sense that the front door is locked into the information we need, so we'll use the back doors into other systems." Jakes fingers typed quickly and the home screens from different drug companies appeared onto his screen. He would try different things to get inside their systems. If it did not work, he try something else. After a few minutes he moved onto the next site. After a half hour he still had nothing. No way to get in and no response to the doctors name with any of the large drug company.

Jake finally said, let's go directly to Tygart University's website and see what we get, if we type in, BIO-GENETICS RESEARCH." In a few moments the screen listed the different departments along with the names associated with areas of research.

Charlie looked at the screen, "nothing, no name, no indication she was even there. Surely you can find some information if she got any federal money under the Freedom of Information Act?"

Jake said, "Relax Charlie, don't look for what you see, look for what you don't see. And I don't see any departments or monies associated with genetic research at the university. Any major university worth its salt is going to have some research in the sciences. But, this school doesn't have any showing. That tells me something's been removed, and removed lately and quickly. The IT department has not had a chance to update the screens with new information. It is just easier to wipe a screen and reinsert a new one than to go into the old and make corrections. Saves time. Let me show you something."

Jake started making simple information inquiries. He sent dozens at a time. Then dozens more. He made it look as if the inquiries were coming from different computers all around the country. "I wrote this program about ten years ago. Ask a simple question of the target computer, like what are your business hours. Then have the question routed to different libraries around the country. The host computer tries to process the mass of questions. It does not equate the information requested as a red flag. No security firewalls have been breached or even bumped into. I have simply asked, what are your business hours? The question is being asked a thousand times in a few seconds. Simple request from hundreds of library computer systems come into the servers all at once. And the university systems don't see the information request as a problem, but it is not able to handle the volume of request."

Charlie could see the response time was slowing down as the host system tried to keep up with the information request. Suddenly the college's home screen froze and a message appeared. HOST COMPUTER NOT RESPONDING. TRY INQUIRY LATER.

Jake looked up at Charlie, "We did it. Now we wait."

Charlie looked at Jake, "We did what? Looks to me like we crashed the college's damned computer. We are further away from the information than we were before."

Jake smiles, "Maybe, if we are lucky the computer got mixed up and ran home to the main server. It will reboot itself from the last good entry. Maybe, just maybe, they didn't think to erase the backup information in the event of a crash. They will do it the next time they do a complete reboot of the system. I am just hoping they have not done it yet."

Jake typed in an inquiry to the university's genetics department. He pushed enter and a circle appeared on the screen indicating the computer was working.

Jake pointed toward the indicator, "This computer will stay attached and keep requesting information until it gets what it wants, or I stop it. This may take 10 minutes or hours. We'll see what happens. You want something stronger than that coffee?"

"Yea," Charlie said.

For the next two hours the men sat and talked of the old days. They laughed and talked of cases they had worked on in the court. All the while both men watched the indicator going in endless circles.

Then as the two men were running low on old stories the screen came alive. "WELCOME TO TYGART UNIVERSITY BIO-GENETICS DEPARTMENT."

"Hot damn, we're in! What was that woman's name?"

"Elizabeth Mathenson." Charlie answered quickly.

Jake typed in the name. In a moment a new screen appeared. On the left side was a picture of the doctor. On the right side it told of the advances in bio-genetic medicine her department had made. She had won dozens of awards and was considered one of the top people in her field worldwide. Toward the bottom of the screen Charlie saw something of interest.

"What's that" he asked.

Jake moved the cursor over the link and tapped it. A new screen appeared. "Mammoth Cloning Project."

Jake looked at it for a moment, "This isn't what we're looking for, we can come back later. I want to go back to the other screen and see if we can get more information about her grants or corporate sponsors."

The original home screen appeared once again and Jake was going to different parts of the home page, looking for information. It seems to be the kind of information that you could find on any home page, pretty generic stuff. Lots of fancy words with professional pictures, but not much of substance. Then the screen suddenly turned blue. "WEB PAGE NO LONGER AVAILABLE."

Jake sat back into his chair, "Wooo… that was fast. I didn't think they would discover the reboot so quickly. Sorry, but the door is closed. I'll keep trying other sites. They can't have hidden them all, sooner or later I'll find a way in."

At the bottom of Jakes screen a picture of a Goblin face appeared with the word, WARNING just below it.

"Shit, their on to me!" He quickly typed a couple of keys and the warning vanished. "Horseshit Charlie, that was close."

Charlie looked puzzled for a moment, "what just happened Jake?"

The server's security systems recognized I was looking at their site without authorization, their security sent a computer worm or virus to invade my computer and do to me what I was trying to do to them. But, I am smarter than that, it only got into the quarantined files in my computer. Every time you go to a site, they send what they call cookies to your computer. It is a way they can identify you each time you access their sites. Pretty harmless for the most part. But, I don't think this time it was a welcoming cookie." Jake turned and looked at Charlie, What the hell have you gotten us into?"

Charlie thought about Jake's question before responding. Without answering he said, "Go to the Mammoth Cloning site. You know the one at the bottom of the page, can you go back there?"

"Sure, that is easy." A couple taps of the keys and the site returned.

"Okay Charlie, let's see why this site was listed on the genetics' department web page."

The page reappeared. "A lab in Korea is trying to clone a Woolly Mammoth from tissues samples found in Siberia. I've heard about this stuff. Kinda' like that movie where they created dinosaurs. Instead they are trying to engineer a hairy elephants. Not as scary as a T-Rex, but this will someday come to pass. Kids in the future will go to zoos and see a genetically engineered mammoth. They'll not ask how much of the creature is pure Mammoth or how much is from a modern elephant. The kids will just be amazed. They are using some cutting edge technology to try and do this."

Charlie reads the information on the screen and then asked, "So, you're saying they can do this? They can create some bastardized creature just to say they did it? Another damn golden trout!"

Jake looked at him with a questioning look before he said. "Well, not yet, but they will someday. The genetic information in the mammoth's tissue is badly broken down. They have to splice it with elephants to repair the sequence. Far as I know they have not done it yet. But someday they will. And what's this about a golden trout?"

Charlie again ignored his question and pointed at the area of the screen where an inquiry can be typed. "There, type in Elizabeth Mathenson."

Jake entered the name and in a split second a home page screen appeared. On the page was a picture. In the picture were several people wearing white lab coats. They were all standing around a model of a molecule near the center of the table, along with a microscope. Charlie recognized Elizabeth's picture. "What is a doctor, who is doing genetic research at the university doing in a picture with people trying to bring back a mammoth?"

"Don't know about your doctor, but look at the names and credentials of those people with her. All with doctorates and PHD's out the ass. This appears to be a gathering of the top people in the field of bio-genetic research, and the judge's old girl friend is right in among them. Don't know what that means, but she is an important part of the genetics' community. That raises the question, why would the university totally purge a world renowned doctor from the school's web pages. This kind of person brings in big grant money. You know how many people want to clone their dogs? I'm guessing there are some people that even want to clone a kid. This is becoming more interesting already. Trouble is, it don't get us any closer to the money question than we were before. But, one thing is for sure, this program is big, really big."

Charlie sat down and processed the information he knew so far. He thought out loud, "A lot of information about what she is working on. Could the money have gone into cloning a Woolly Mammoth? The judge said when he knew her, she was working on genetic research for diseases. Could she intentionally shifted that research toward the mammoth and away from decades of bio-genetics' disease research? If that's what she did, I guess the government and the drug company's would be pissed about that. I'm guessing some government audit could have turned up that she was spending tax payer money on a hairy elephant project in Korea. That would get the media's attention real quick, that's for sure."

Charlie looked at the screen and the information listed below the picture. He pointed towards Mathenson's name. Go to the list below her name. There is a link, see where it takes us.

Jake touched the link. Instantly the screen changed to a new home page. "Wah-Whoooo!" Jake called out. "Were in."

The page was titled, The Amino Dimatrix Atom Molecule Project; by Dr. Elizabeth Mathenson.

"What the hell is it?" Charlie asked as Jake leans close to the computer screen and begins reading the information on the home page."

Jake said nothing for a few moments as he continued to read. He then moved on to another page listed on the site and studied the information listed. It was hard for Charlie to see the words from where he was positioned, looking from the side of the screen, but he could clearly see the pictures were of chemical elements and complex molecule structures. He could also see words in bigger font. Words like, DNA, GENETIC SEQUENCING, NANO-MOLECULE, PROGRAMED 3-D PROTEIN MATRIX, LIGHT-SPECTRUM ANALYZED ENZYME.

"Well, what is she into?"

Jake sat back in his chair and scrolled through the pages of the site. "Well Charlie, I can't say I understand all this stuff, but it looks like she was working to change DNA structures with something called the A.D.A.M. Protein Molecule. It is pretty heavy stuff and I have no background in this kind of information. Hell, I don't know how any government accountant could tell if this is what the grant money was to be spent for or not. There has to be a lot more to the story. And this doesn't get us any closer to the story of the money she is charged with embezzling. But, we now know she is a very important person in her field of science.

The people who build these sites tend to make them bigger and bolder than the subject they are promoting. Think of this page as an advertisement for donations of money to get a piece of the action. Drug companies want the science of a potential new drug technology. They are hoping it is the key to developing the next generation of genetic based drugs. If they can make this work, it is worth millions, no I'm wrong, this is worth billions with a capital "B."

Jake was looking over the information, and as he did he said to Charlie, "I know a don at Stonemore University in England."

Charlie got an odd look on his face at the term, "Dons".

Jake went on to explain, "Here in the states we call them professors, but over in England they are called, Don, I think it means old world Lord. You know the English, they have some odd titles for things. Anyway, he's big into using light waves to analyze chemical elements. He had an idea if he could solve some problems, he could map the human genome with computers and greatly reduce the time needed. He had some idea if he could map the genome using the light waves, much as he could the light waves from outer space, then he could program a modified computer program to mathematically recreate the sequencing of the DNA structure. He even believed if he could, it might lead to creating synthetic DNA. Problem was, he could not find a computer to do the complex calculations. Even Stonemore's massive research computer system could not dedicate enough time to do the sequencing. The last time we talked shop, he said it would take Stonemore's computer working 24 hours a day for two months to do the calculations. That's one big mathematical problem. That was a couple years ago, but I'll contact him and see what he can tell me about this information and where the doctor would have gotten her funding. He may know what she was into. If this lady was into this type of research, she had to also talk with the people at Stonemore. They live for this kind of thing. They also believe the question of life is found in the words of the Bible. Hard to figure they are working the age old problem from both ends. In the meantime I suggest you go see this doctor and find out for sure what this is all about. Remember, from what we know about the indictment, this is only about the money and not the work she was doing. Hell, she might have bought a big house in the Bahamas and vacations there ten months a year, but somehow looking at this information I don't think that's how this lady spent the money. Go ask her."

Charlie turned the screen toward himself to better look over the information. "It says here she is on the board of the National Cancer Center. This lady is well thought of and has been part of the research community for a lot of years." He turned the screen back towards Jake. "I have to remember this is about the money and just the money. Maybe she'll give me some insight into the indictments. That's the only way I am going to know if I can defend her."

Jake sat staring at the screen. After a minute he said without looking at Charlie, "You know this might be something else. What if… well let suppose you could clone your wife or son? Would you want to? This stuff is all about dogs and sheep and mammoths, but if they can do those animals, what stops them from cloning people? Not saying they are, just saying it is not too great a leap from a sheep to a kid, when you're talking genetic cloning. You can tell I watch a lot of the science channel."

For a few minutes both men sat thinking about what they had learned so far. Finally Jake said, "Charlie, you need to go and see her lawyer, and see if she even wants our help. If she don't, this is purely a waste of time for you and me. Get me a copy of the indictment and fax it to me. It will have the information I need to look into the records. You talk with her and find out what is really going on. She is a respected doctor and the university has cut all ties with her. Something has spooked them big time."

Charlie looked at his watch. It indicated it was three fifteen. He calculates in his head for a few minutes. "It will take me a couple hours to get to Monongahela City. By the time I find the lawyer's office he will be closed for the day, most likely. How about I stay here tonight and get an early start in the morning?"

Jake smiled real big, "I'd like that. Been cooking for myself for the last year. Be good to cook dinner for someone besides myself."

Charlie grinned at him, "Well, I planned on taking you out to the steak house for a big steak and a couple of cold beers, but if you would rather stay here and cook, that will be fine."

Jake jumped up and headed into the house. "I like my steak rare and a dark beer. Give me a few minutes to clean up and we'll be on our way. I'll never turn down a free steak and cold beer, not for some damned crappie fish." The two men laughed.

Charlie turned the screen of the computer so he could look at the information while he waited for Jake to return. As he examined the picture of Dr. Mathenson, he said out loud, “Lady, what the hell have you got yourself into?” He paused and sat back, “Charlie Morgan, what the hell have you gotten yourself into?”

Chapter 7
The New Face of the Law

The next morning, Charlie and Jake said their good-byes over a cup of coffee before he left for Monongahela City. Last evening over dinner they talked of past embezzlement cases they had worked on together. Jake explained to him that with a computer and the ability to do your banking worldwide, sometimes it is hard to track money of those who know how to hide it. But, they did not think this was the case. They both agreed the meeting with the lawyer would help them find out basic questions about the coming indictment of the doctor, or end their involvement all together if she was guilty of simple fraud and embezzlement.

It was near eleven o'clock when Charlie made it into Monongahela City. He hated driving in the traffic of a college town. He figured young college kids were always in a hurry to get somewhere. They had not yet learned to slow down and enjoy the view. Of course Charlie remembered he had not learned that lesson himself, until he was in his late 50s.

He glanced down at the slip of paper lying beside him on the seat with the lawyer's name and information. He craned his neck looking for street signs. Finally, he saw a marker with the name, Locust Street. He said to himself as he looked at building numbers, "Now, Charlie my boy, that lawyer's office should be somewhere on this block." Charlie often talked to himself, sometimes he even argued with himself.

As he drove slowly along the busy street, he soon figured it would be easier to park his truck and continue his search on foot. Luckily a delivery van was pulling away from a spot on the corner. He quickly pulled in the opening on the busy street.

Charlie parked and began walking looking for the address and the lawyer's office. It was an old section of town and the building numbers were not always in place. Finally he found what had once been an old brownstone house. It looks as if it had been changed into offices several years ago. The building was dirty and the window frames needed a couple coats of paint. In fact, it looks like the building's exterior needed a lot of work, he wondered about the inside. He would soon find out.

He opened the door into a room where a secretary's desk sat. From the looks of it, no secretary had yet called it their own. The desk looked as if it was being used as a temporary file cabinet. Along the back wall were unopened boxes. The writing on the outside of them indicated they were mostly law books. It also looked as if they had not been used yet.

He looked about the office and saw no evidence of anyone being around. So he decided to call out. "HELLO, anyone here?"

For a few seconds there was only silence. Then he heard a sound from the back, as if someone had tripped over something. "Who's there, just a minute, I'll be out."

Charlie stood looking around as he heard footsteps coming his way. The footsteps belong to a tall young man. His sandy hair was a bit messed up and his shirt tail was partly pulled out. The tie around his neck had been tucked into his shirt. As he approached Charlie, he wiped his hands on a paper towel. Once they were dry he offered his hand to Charlie.

"Good morning sir, please excuse the mess, I never realized getting an office in order could be so dirty. "My name is James Downing, may I help you?"

Charlie stumbled for a moment trying to figure how to get this conversation off on the right foot. "Well, Mr. Downing, I guess I am here to see if I can help you. I have been retired for a few years and I wanted to get back into the courtroom, not so much as the front man, but more as a research assistant. I heard you have a case of embezzlement concerning a doctor at the university. I use to specialize in embezzlement cases."

The young man was surprised by Charlie's offer and suddenly a little suspicious of the man who walked in off the street and referred to his case. Especially this case.

"Well sir, if you'll excuse my curiosity, I am a little surprised why you're asking about that particular case and, if I may ask, just who are you?"

Charlie realized he had forgotten to introduce himself. He knew in the back of his mind he must be a little rusty in his opening presentation. Charlie's first rule of fact finding, never give the other person the reason to ask the first serious question in a conversation. Gives them the edge right off the bat.

"My name is Charles Morgan, but most people call me Charlie. I am a federal prosecutor, or was, I am retired now and spending time fishing." Charlie knew the fishing part would throw the young man off his thought pattern for a few moments.

The young man adjusted his tie and tucked in his shirt. He may have been new to the law business, but he adjusted quickly to the situation. He took on the demeanor of a determined lawyer.

He completely ignored Charlie's comment about fishing as he asked, "Mr. Morgan, were you once the federal prosecutor from the northern district working mostly in Elkins?"

Charlie shook his head as he said, "Yes that was me a few years ago. Last few years I been a fisherman and a guy who sat on the porch and watched the neighbor's dog try and catch passing cars."

Without any further discussion the young lawyer said, "Sorry, Mr. Morgan, I don't need your help. I am surprised that the government would send someone like you to spy on me before the indictment has even been served. Now, if you would excuse me, I have some unpacking to do." With that the young man disappeared down the hallway.

Charlie thought for a moment as to what he might do next. He figured as quickly as this young man dismissed him, it was unlikely he would change his mind before he threw him out of his office. He had maybe one chance.

"Mr. Downing…James, I have a message for Elizabeth."

He waited. After a few moments he heard the footsteps returning. The young man came to where he could see Charlie.

"Well, what is your message?"

"Tell Dr. Mathenson I came by and tell her the summer of 81 was a very good year."

Downing came a little closer as he said, "That's it…you just want me to say the summer of 81 was a good year?"

Charlie realized he had gained control of the situation once again as he said. "Yea, that should do it. And don't forget to tell her Charlie Morgan delivered the message."

Charlie knew Mathenson did not know him, but if she looked his name up on the internet she would find out he was a prosecutor in Peterson's judicial district. He hoped she could understand the message. He didn't have any other options at this point. She just had to make the connection on her own if she wanted his help.

Charlie looked over on the desk and saw a pen and a piece of paper. He picked up the pen and wrote something on the slip of paper. He turned to the young man and handed it to him. "That's my cell phone number. I'll be in town until the day after tomorrow. If I don't hear from Mathenson or you, I'm heading back to the mountains to go fishing. Do you understand Mr. Downing?"

The young man looked at the paper and then to Charlie. "That's it, a phone number and the summer of 81? Nothing more?"

Charlie started toward the door once again. He turned to Downing, "Son when you have spent more time in a court of law, you'll know it is best when putting up a good defense, the less said the better. Don't give the others side more than they already have." With that Charlie stepped outside and looked toward his truck. He figured he needed to find a hotel room for a couple of nights. He also remembered an old friend in the law department at the university. It had been nearly 15 years, but Charlie hoped he was still teaching. If he was, maybe he could give him an idea of what was going on. He looked at his watch and saw it was one o'clock. He figured the man he wanted to talk with might be in his office in the afternoon. Charlie decided to stop by the law department before finding a room for the night.

As he headed toward the law department he found himself lost in the streets of Monongahela City. He had not been in the college town for several years and much had changed. After driving around he finally decided to ask someone for directions.

At the corner of the next block, two young men wearing backpacks stood looking at their phones. Each of them wore baggy shorts and earphones. They both seemed to be listening to music and not the clamored sounds of the street. Charlie stopped and leaned out the window and motioned to the two young men as he said, "Excuse me gentleman, can you direct me towards the law center?" At first the two looked at him like he had spoken some foreign language. Finally one of the shaggy haired boys pointed down the street and said, "See the second stop sign mister? Well, you go down there and turn left, then go to the top of the hill, you can't miss it." Charlie thanked the young man for his help and headed toward the distant stop sign.

Sure enough the boy was right, Charlie found the building and headed inside. A guard sat reading a magazine. He never even looked up as Charlie came into the building.

"Excuse me, excuse me." Finally the man looked up at Charlie.

"Yea, what can I help you with mister?"

Charlie asked, "Can you call Professor Gordon and ask if he has time to speak with Charlie Morgan?"

The seemingly disinterested guard picked up the phone and made the call. Charlie could hear the phone ringing. Finally, someone answered. "Yes, what do you need?" the voice asked.

"A man named Morgan, what did you say your first name was?"

"Charlie." He responded.

The person on the other end must have heard Charlie say his name and told the guard, "Send him up to room 211."

The disinterested guard said as he returned his attention to the magazine, "He said…"

Charlie interrupted, "I know, 211."

In a few minutes Charlie got off of the elevator and looked at the sign on the wall. Rooms 201 to 210 to the left. Rooms 211 to 220 to the right. Charlie headed down the hallway toward the room.

He opened the door slowly and looked inside. He was met by a distinguished man who looked to be in his 70s. "Charlie, Charlie Morgan. I thought you were dead. Haven't heard anything about you in several years. It used to be every high profile case that needed prosecuting in the east was handled by you. I figured you retired or died when I hadn't heard from you for so long. Come in, please come in. Tell me all about your adventures since you gave up putting white collar criminals away."

Charlie looked toward the back of the room and saw two students working on papers. He looked at the professor. "Clay, can we have some privacy before we talk?"

Professor Gordon looked at Charlie and slapped him on the arm, "Of course we can. I see this is not purely a social call, something important on your mind?"

Charlie shook his head, confirming the professor's thoughts. Before long the two students had gathered their books and were leaving the class room. As they did, Clay closed the door behind them and offered Charlie a chair.

"What's it been, 10, 12 years since we last talked? You know, you still own me a guest lecture to my class. These kids today need to hear from an old prosecutor about the law in the big court rooms. Hell, they all think it's like they see on TV, suits, ties, big cars and women. They don't know how hard you have to work to put together a big case. But, you could tell them how it works in the real world."

Charlie sat back in his chair as he said, "Clay, the world of law has changed. What we knew about it 40 years ago has changed a lot. These kids have a hard row to hoe if they think they'll walk out of here and get a position with a firm making big money. Hell, they'll have a hard time just getting billable hours the first few years. I'd hate to be starting out all over again."

The two men talked for nearly an hour on everything from changes in the law to some of the decisions the Supreme Court had handed down.

Finally, after a while, Clay asked. "Charlie, I know after all this time you did not come back for this conversation, even though I have enjoyed the hell out of talking with you. What do you want?"

Charlie leaned forward and in a hushed tone said, "I want to know why the university cut Elizabeth Mathenson lose?"

Clay never changed the expression on his face, "Charlie, how the hell are you involved in that? People are not even supposed to know about what happened, they, the human resources people have kept it quiet. HR let it leak out that even those of us who have long tenure should not talk about the case."

Neither man said anything for a few minutes, then Charlie said, "I figured I'd try and get a handle on what was going on before I decided to get involved. And since I have no one else to talk with, I came to you."

Clay smiled at Charlie, "When can I expect you to come and speak with my classes for one week?"

Charlie realized he was being blackmailed, but he saw no alternative to getting the information he needed. "I'll see if after mid-terms I can make it up here for a week. One way or another I will speak to your classes. Send my secretary a note at my office. You just contact her and let her know what I agreed to do."

Clay slapped Charlie on the knee, "Good, I knew our meeting to discuss your being a guest lecturer would pay off. Now, to the other question, truth is I don't know, well at least I don't know anything official. But, rumors are she got caught using the federal grant money for unethical or illegal biological genetic research. Something called the ADAM Project. She has made several trips to Korea in the last year, four to Africa and three to Israel. Furthermore she spent two weeks in Edinburgh Scotland. She was only supposed to be there for three days. I also heard she was in England at Stonemore for a period of time. Far as I know she never gave any explanation of where she was for the rest of her time. Whatever happened, the State Department contacted the university higher ups. Not sure what they said, the only thing I know is after they talked, Elizabeth was out, and I do mean out. They cut her lose with no pension or severance pay or anything. They just came in and told her to pack and get out. She threatened to sue the university for unjustified termination and breach of her tenure. The next day the rumor spread she was being investigated by the FBI and most likely was going to be indicted by a federal grand jury. That's about as much as I know. Whatever she did, it pissed a lot of people off and they are willing to send her to jail. And it all has something to do with the federal grant money used in her projects. Her department has been shut down and locked up. Elizabeth has been around for a very long time, nearly 40 years. Her program has more doctoral applications than any other research program the college offers. It is rumored she had some very big pharmaceutical companies backing her research. It was also rumored they were highly pissed that the university locked up her program and research records. And when I say locked up, I don't mean they just locked the doors, no... they have armed guards with sidearm weapons at each door and the building is under the watch of new security cameras around the perimeter. The guards wear no ID and offer no conversation when approached. Those who do try and talk with them are soon taken away by campus police. Not sure what happens next to them. One staff member said they took a couple of kids down to the local police station and kept them up all night asking what they were doing at the building. After about 24 hours, they released them with no explanation. That arrest pretty much keeps others away. Even before this, Elizabeth had one of the largest departments and one of the most secretive departments on campus. It took an act of God to even get inside the building. More security than Fort Knox. Her knowledge of biological genetics for research brought in big money from the government, drug companies and private donors. It is also rumored she received support and money from foreign governments who don't like oversight. Our state has vague laws on the books governing bio-genetic research. A recent report stated ours is one of only a few states where medical research is like being in the wild west, few laws. I think Elizabeth's department may have taken advantage of that lack of oversight. Anyway…the people who did come and go from her department were the secretive types, they traveled with security people. You know the kind, they wears sunglasses even at night and have ear communications devises like the secret service. When they came onto the campus, everyone gave them room. I don't know if you realize it in your line of work, but medicine in the future will be more about genetic engineering than scalpels and drugs. Whatever she was working on, it drew a lot of interest and very little information went outside her lab. I think maybe she got in over her head and what she was developing became so valuable or important, it took on a life of its own. She may have become a victim of her own success."

Charlie adsorbed the information and then asked, "Do you think she was into something she shouldn't have been."

"Charlie, I teach law, I am not going to speculate on anything. All I can say is I have been around the school since I came to this country, nearly 40 years ago. I have never experienced or seen anything like this. Last summer after a faculty meeting with the dean, she took me aside and asked if there was any way around the university taking control of a potential patented research project. She knows that I have a great deal of experience in international patent laws."

In the hallway just outside the door Clay heard footsteps that stop by his door. In the frosted glass he could see the silhouette of a person. Suddenly Clay stood and said, "Charlie, I don't want to be a bad host, God knows I have enjoyed seeing you today, but you have to go."

He leaned close to Charlie and said in a soft tone. "For the record, you came by to tell me you are going to speak at my federal law class just after mid-term. That's all we talked about. I'll walk you out. After what they did to Elizabeth, I don't want to put my career in jeopardy, tenure did not save her, and she had a lot more clout around here than I do."

Charlie somehow knew his old friend told him everything he wanted to tell. But, something in his look and tone of voice, Charlie knew there was more. Clay was a lawyer that meant he was supposed to be able to hide his thoughts and emotions in a court of law. But he had been too long away from the courtroom, Charlie saw something in his face. It was also clear Clay did not want anyone else to know he had talked with Charlie, other than to ask to be a guest lecturer. He noticed something else that made him wonder. He saw on the table behind Clay's desk only two pictures. One of the pictures showed four young men in military uniforms. Clay was one of them. In the background was an Israeli flag. The second picture, more up to date, was of Clay standing alongside the Prime Minister of Israel. Clay must have some important friends.

The person outside had vanished as they started for the door, Charlie stopped and asked one last question. "She's got a young kid for a lawyer, what do you know about him?"

Clay smiled, "Good kid, smart as they come in understanding the law. He was particularly good at seeing the underlying meaning of the law. But, still as green as grass when it comes to real world law. Had him in several of my classes. I'd have to say, he has a head for the law and someday I'm bettin' he'll be one hell of a trial lawyer. But, in this case, he'll be in way over his head if he tries to defend her on federal embezzlement charges of this complexity. He doesn't have the resources to investigate the scope of the government's indictment. What I can't figure out is why Elizabeth would go to him for representation in the first place. The only reason I can figure is, a couple of semesters back, the kid wrote a paper on new genetic medicine and its legal standings. He spent some time in the genetics lab talking with some of the researchers. Don't know if he talked with Elizabeth, but I'm figuring he might have. Elizabeth's best option, try and plea bargain it down to a lesser charge and get a seasoned lawyer, a lawyer like you. I'm guessing no one is wanting this case to come to trial. Too hush hush already. Elizabeth should have stayed with her original research and kept her nose out of other country's businesses. One other thing you should know. It has been said that after she left computer files and a cryogenic container were found to be missing. They suspect she knows something about where they are."

Charlie wondered about Clay's final comment, but did not ask any further questions. He suspected he had gotten all the information he was going to get from Clay. He shook his old friend's hand and turned and walked away. Charlie had known Clay for many years and considered him a good friend, but somehow their meeting today gave him insight into the man that he had never known. He then stopped and turned back to Clay, "one more thing, do they still have programs to get people into the law school that have real world work experience?"

Clay looked at Charlie with a bit of an odd look, "Do you mean for people who have college degrees and could not meet the guidelines for the program at the time?"

"Yea, something like that. They both work for the Justice Department as U.S. Marshals. They would like to get their law degrees. The Justice Department sometimes helps with arranging schedules and finances. I would make a recommendation to the department, could that get someone into the program?"

Clay shook his head yes, "Do you have someone in mind?"

Charlie shook his head indicating he did. Then he said, "I get through this Mathenson thing and I'll get back with you on a recommendation I would like for you to make for me."

Clay waved as he said, "Shalom my friend."

With that, Charlie turned and started toward the steps on his way down to the main floor. It gave him a few minutes to think about what Clay had told him. The one thing he was sure of, Clay didn't know much about what was going on, that was until he started talking. And why make a comment about Elizabeth staying with her original research program and keeping her nose out of other country's business. And he indicated Elizabeth was asking about patents and ownership. He knew anyone who works for the university knew that any discoveries would be the property of the university. Why ask about patent rights? Could Elizabeth's research, which is being funded by the government and private drug companies, somehow become in dispute with the university? Charlie remembered Clay was well versed in patent law especially international patents. He often used his students to work with him to research patent rights and ownerships. Often they traveled abroad to work on summer projects. And most curious of all, Clay's comment about the indictment. Charlie figured it was coming, but it had not yet been released. Charlie wondered about his old friend and what he may have been hiding. His courtroom sense told him Clay somehow had a hand in this and he did not want Charlie to know, or did he? He began to wonder, was he the one being used in this game of misinformation? Charlie made a mental note for Jake to look into Clay's background, especially his involvement with the military and government.

Chapter 8
Doctor Mathenson

Charlie's neck and back felt tired as he came out to his truck. He looked at his watch, it indicated quarter-until-four. He thought to himself, had I been talking to Clay that long? He also realized he had not eaten since early that morning and his blood sugar was low. It was time to get a room for the night and something to eat.

He remembered the lodge out at the lake. It served a good steak and the rooms were nice and overlooked the lake. A pleasant, peaceful view in the morning to sit and drink his coffee. It would take him 20 minutes to get there, but he didn't mind the drive. Gave him more time to think.

Before long Charlie arrived at the hotel and secured himself a room for the night. Wanting to unwind, he stood in the shower letting the water run over his tired bones. He said to himself under his breath, "Charlie Morgan… you're too old for this shit." As he dried himself he watched the Weather Channel. He had turned down the voices and was just watching the pictures. He hated listening to those weather people talk endlessly about the weather in someone else's back yard. He always thought to himself, "Just tell me the weather today, I don't give a shit about yesterday or this morning's weather in Dallas, Texas. About then his phone rang. He looked at the incoming number, he did not recognize it. He answered it, "Charlie here, how did you get this number?"

There was a momentary hesitation on the other end. Then a soft voice said, "Mr. Morgan, sorry to bother you, but I think you wanted to speak with me… you know…about the summer of 81."

Charlie sat down on the edge of the bed as he thought about his next words. "Doctor Mathenson, I presume you know who I am by now. Either your lawyer has told you of my background or you have looked me up on the internet. Either way, I think we need to talk."

"My lawyer says I should not talk with you. It could be a trick by the government, after all you were a federal prosecutor for many years."

Charlie sat back on the bed and sighed deeply. "Listen Elizabeth, two days ago I was in a place where there was just me and God. And the truth is, I felt crowded. I was trying to catch a fish that I have pursued for years. It is a place where I hear only the rushing water and the sounds of eagles from time to time. No phones, cars, loud kids or annoying people. But, I was dragged out of my sanctuary and asked if I could help you. I don't know you or your situation and quite frankly, I'm not sure I want to become involved in your circumstance. But, I agreed to offer my help. You give the word and I'll be back in Smoke Hole Canyon before you know it. What's it going to be?"

Again there was silence from the other end, until, "Ok Mr. Morgan, when can you meet with me and my lawyer?"

Charlie thought for a moment, "Do you like prime steak? The lake lodge serves one of the very best around. Join me for dinner and we can have a friendly off the record conversation. I need to know what I am getting into and you need to know if you want me here, is it a deal? Two people having dinner and deciding if that is all it will be. You tell me to go and I'll forget everything we talk about."

The soft voice said, "Deal. My lawyer won't like it, but he never said not to have dinner with you. I'll see you in an hour in the dining room. Ask for the table by the window on the south end. The wall partition blocks the view from the main dining room." Charlie agreed.

After a half hour, Charlie made his way to the dining room and was able to arrange to have the table Elizabeth suggested.

He ordered a whiskey on the rocks and sat back watching the sun glimmer off the late evening lake. He thought of his fishing place and wondered if The General missed him.

As he sat there his mind began to wander, and for some reason he thought of his wife and son. He missed them as the suns reflection glimmered off his face.

"Hello, are you Mr. Morgan?" A gentle voice asked."

Charlie looked up, the reflective evening light illuminated the face of an attractive woman. She wore heavy dark rimmed glasses. Charlie had a quick flash of a Clark Kent's image. Her silver gray hair was neatly pulled into a twist on the back of her head. She wore a white blouse and dark slacks. Somehow she did not look like a stuffy old research doctor. Charlie quickly remembered his manners not to prejudge as he stood and offered her a chair.

"Yes…I am Charlie Morgan. I assume you are Doctor Elizabeth Mathenson?"

"Yes, that's me, and please, call me Elizabeth. I see no need for formalities."

Charlie smiled as he agreed, "Yes, and please call me Charlie. Won't you please sit? Can I get you something to drink?"

Elizabeth sat down and looked up at the waiter, "Please, can you get me a vodka tonic, heavy on the vodka?"

The waiter indicated that he could, he then asked Charlie, "Sir, may I get you another?" Charlie shook his head, indicating yes.

Elizabeth looked out over the lake, "I have forgotten how beautiful it is at this time of the evening along the lake. I use to come here often, but not so much anymore. How about you?"

Charlie looked at the lake as he said, "Not for a good while. I use to come here in the past with my wife, but that has been several years ago. Haven't made it back for a long while."

"Is your wife not with you this trip?" Elizabeth asked.

Charlie hesitated for a moment, "My wife was killed along with my son a while back. Since they died, I have not been much on socializing. I spend much of my time at home. On occasion I like to get back to the mountains and try to catch one special trout. I call him 'The General', my wife always got a kick out of me naming a fish."

Elizabeth got a questioning look on her face. "You said you were fishing a couple of days ago for a fish that you have pursued for several years, even as far back as when your wife was still alive? Excuse me for saying that I don't think hatchery raised trout live that long in the wild. Is there something special about your particular fish?"

Charlie was not sure how to answer her question. "Well, I never really gave any thought to how long a fish may live. Maybe you are correct, they may not live all that long. The fish and I have this thing where I try and catch him. If we never really meet, I guess I'll never know if he is as old as I want to believe him to be. I know it sounds a little weird for an old lawyer to be on a mission to catch a particular fish, but it may be the one thing I can look forward to in my life these days that brings me some joy."

Elizabeth pursed her lips for a moment, "I am sorry Charlie, I did not want to imply you are chasing a simple every day fish. If I remember correctly, pond raised trout live for only a short time in the wild. They are genetically grown to mature in size quickly, survival in the wild is difficult for them after being raised in concrete ponds. It sounds like you and your fish have been in this struggle for several years."

Charlie responded, "I believe "The General" is a native rainbow trout born in the waters of the Potomac River. Hatchery fish are engineered to grow quickly and not necessarily to survive in the wild. But the worst of them are those bastardized golden trout."

Elizabeth sensing the importance of Charlie's fish said, "I think you are right, most likely yours is a native trout and would live much longer. Natural selection I'm betting." She wasn't going to ask about his comment about golden trout.

Charlie smiled and the two sat saying nothing for a few moments. About then the waiter returned and took their orders. After that they spoke of things in general. But, by the time their dinners arrived they had talked about many things, everything but the topic they were to discuss after dinner.

'Charlie, you seem like a good man and not the enemy that my lawyer told me to watch out for. But, you have to know why I am suspicious of your sudden involvement. I am going to assume you were asked to help because of someone I knew a long time ago?"

Charlie thought for a moment. "That person from your past could not get involved in anyway. In fact, if your case comes to trial in the Elkins district, he will be playing golf in Myrtle Beach. If he just happens to be away recovering from a stressful year, no one will think anything about him being gone. He is an ethical man and knows that is the only way. He wants an old friend to be sure and get a fair trial, for whatever is going on. And quite frankly, I don't know what is going on. And can you tell me why you retained a lawyer who has never stood before a jury?

Elizabeth took a sip of her drink. "Well Charlie, the truth is pretty much as you have been told. The government auditors and the college have accused me of embezzling project funds. They believe the research my team is working on is not contained in the original grant program goals. They believe my research should blindly go only in one direction. Years of research have led me to believe there are several possibilities for the answers to life's big questions. Recent discoveries by my colleagues around the world have shown me a new path to the truth in my genetic research."

Charlie sat and listened. When she finished he asked, "What about all those trips to Korea, Africa, Israel, Scotland and England?"

Elizabeth looked a little shocked. "Well, I am surprised you know my research is into genetic medicine. The Koreans are doing major research into genetic manipulations of genes. They have found ways to repair damaged DNA strains. Their research is funded by the Korean and Russian governments. They have greater latitude in the direction of their research. As you may know, they hope to create a modern version of the Woolly Mammoth. Regenerating damaged genes is an important part of my work. Do you know that many of the cancers are caused by damaged genes from both a person's family history and by environmental factors? If the Koreans can repair damaged genes in a ten-thousand-year-old mammoth, maybe the same technique can repair a damaged gene in a small child with genetic leukemia."

Charlie thought for a few minutes, absorbing her words. "Ok, I understand Korea, but what about Africa, Israel, Scotland and England? Are they hot beds of genetic research?"

Elizabeth sat back and clearly thought about her next words to Charlie, "Do you know anything about the origin of man?"

Charlie finished his drink and shook the ice in his glass to get the waiters attention. He then smiled as he looked Elizabeth directly in her eyes, "Now Elizabeth, you are asking a lawyer a trick question. Do you mean, where Charles Darwin said we came from or where the Bible says we came from?"

"Both," Elizabeth says with a bit of excitement in her voice.

"What do you mean, both?" he asked with a bit of a disbelieving look.

"Charlie, in the Bible it tells us that man was created by God and lived in Eden with Eve until they were cast out."

Charlie frowned, "Elizabeth, your embezzlement case has nothing to do with where the Bible says we came from. It's about the money you spent and how your program spent it. Open and shut. Believe me, I prosecuted many of these cases. You go on the witness stand talking about Adam and Eve and that apple tree, you're in big trouble"

Elizabeth leaned toward Charlie as she said, "Okay, but just hear me out before you go making any quick decisions. The Bible says we came from two people in the Garden of Eden. Science say mankind first appeared somewhere in northern Africa. You see, Charlie, science scholars say the cradle of mankind was somewhere in what we know today as Iraq. Eden is believed to be in that same general area. Both places are in the Middle East. The big difference is the time element. Darwin said man evolved over hundreds of thousands of years. The Bible says man began around six thousand years ago. The big difference is time and how the story is told between religion and science."

Charlie shook his head, "Elizabeth, again I have to tell you this is not about the origin of man, it is about how your research program spent the money. The government is not going to get into your theories. It will tell the jury about the money they gave you and then tell them you spent it on something else. The prosecution will narrowly define the money's purpose. Then they will ask 12 men and women to decide on those narrowly defined parameters. Did she spend it as the original grant money was requested? They don't care about where man came from or how long ago he showed up, in either Darwin's story or Genesis's story. Believe me, if that's your defense, you are going to jail if I were the prosecutor of the case. Besides, no attorney is going to allow a simple case of embezzling government funds to become a case of science verses religion. That is a no winner."

"No, Charlie, you are missing the point. Mankind's roots in both science and religion are in the Middle East. If I can find the original gene sequence, my team may be able to find the cures for many of mankind's diseases. We believe the answer is contained in original DNA. Do you known Israel has collected millions of DNA samples? They believe they may be able to trace their peoples back to Abraham. At first, I tried finding DNA samples in the general population of northern Africa near the Rift Valley. Then, unexpectedly I had an opportunity to obtain DNA samples from another source; I took it."

Charlie ignored her comment. He was not interested in DNA or Abraham. He was becoming frustrated with Elizabeth's conversation about the origin of man. He kept reminding himself not to be misled from the money. He asked her another question trying to get the conversation back on course. "Why the trips to Scotland? How does that figure into this research of yours? Jesus's people did not come from Scotland. Does all this travel come out of the department's money that's in question?"

Elizabeth picked up her phone and searched the internet. After a moment she handed it to Charlie, "Do you remember when they cloned a sheep? The Scotts were the first to successfully clone a large mammal. And they did not use an egg and sperm of the parents, they used a somatic cell. They worked with it to develop a viable embryo. You see, Charlie, in each cell is the DNA blueprint passed from our parents. In fact, I believe your DNA contains your entire family history from their beginnings. The key is finding a way to unlock the dormant parts of the DNA's genetic structure?"

Charlie's face showed his disinterest in her story. His frustrations with her explanation was making them both uncomfortable.

But, even with his obvious disinterest, she continued her story. "Charlie, it is important you understand some of what I am telling you. You can't separate the money from my work. The scientist in Scotland created life by taking a piece from one animal and a piece from another, combined them into an egg with its DNA removed. Think of the implications. Endangered species in the wild that may die out in the next hundred years can now be saved. We can harvest eggs from one parent, DNA from another and then implant then in a third. Do you know in 30 years the African elephants will be near extinction? Poachers, loss of habitat. We can save them using this science and research like mine."

Charlie looked at the phone's screen that showed a story of the cloned sheep. Not interested, he laid the phone on the table and sat back in his chair and with a disgusted sound said, "Is that what this is really about, cloning? I am sorry, Elizabeth, my head is beginning to hurt. Like I told you earlier, this is about the money, nothing else, just the money. I'm guessing if you spent the government's money on cloning sheep and elephants instead of medical research, they really are going to be pissed. You have to get it into your head, the possible indictment against you is only about money and anything else will be a waste of time when you come before the court. You are looking at a long jail term if they prove their case… and from what you're telling me… you're going to help them lock you away for a long time. I don't know if I want to be part of that. You can lose the case all by yourself with talk of the origins of mankind."

He took his napkin from his lap and wiped his mouth. Then he shook his empty glass to get the waiters attention. Elizabeth could tell he was becoming irritated with her point of view.

Finally, after a few uncomfortable minutes, the waiter returned with new drinks and sat them on the table.

Charlie leaned toward Elizabeth as he said, "Look, I can see you are passionate about your research and I am betting you are really good at it, but I'm a lawyer, not a scientist and if you were my client, I would advise you to try and plea bargain the case. Ask for your job and tenure to be reinstated. And just tell them you became so engrossed in the research you did not know where you were going until the audit. If there is no evidence of spending funds outside your research, they will believe you got lost in the project and had no intent to spend the money differently than for what the grant was originally written. I am assuming a complete accounting of the grant money will not find any money spent on lavish life styles? I am guessing that is not the case here. Plea bargain and get your career back. I bet you're on the verge of an important discovery that will benefit man. Put this behind you and go back to work."

Elizabeth looked at Charlie with a confident look as she said, "Charlie, you have no idea what is possible. Have you ever prosecuted a case and discovered something that changed your life?"

Her comment and question missed Charlie's attention, as he was thinking of something else at that moment. "Elizabeth, can I ask, why did you go to Downing for representation? You had to know he has no experience in the courtroom and you may have big problems. Do you not understand that? You'll be the one paying the price here if you are found guilty, not him."

She sipped her drink before she spoke, "I met Jimmy a while back when he came to visit the science department. At the time he was doing some research work on his senior thesis. He was examining the law in a changing world of science. He was a clever young man and did not come into the project with preconceived notions of what the outcome would be when he finished. He asked questions and listened to answers. He gave good rebuttal when he saw a missing piece of information or disagreed with my ideas. He impressed me for being someone so young and with little or no experience in my field of study. After he graduated he came by my office. I was out of the country at the time, but he left his business card. On the back he wrote, *starting my own practice, if you ever need representation come see me.* I never gave it a thought that I would need his help. Unfortunately I was wrong. So when the university terminated me so abruptly and I was served a warrant to search my office and home for evidence, the only person I could think of was Jimmy. So, I found his office and asked for his help. Truth is, if I had thought about it a little more I would have probably gone somewhere else, and then again I may not have. I guess I saw losing my job as more important than the possible indictment. I may be wrong on all counts."

The conversation for the moment was replaced by the silence of the room. Finally Charlie said, "I believe you have a son, what does he do??"

Elizabeth grinned and chuckled, "Yes, my son is a doctor and works at the university hospital. He's married and they have twins. He has a surgical residency and works twenty-four hour days. I had hoped after he saw how I spent my life working, he would have tried something different. Tells me in a couple more years he wants to start his own practice. Family practitioners are a dying breed these days. I personally hope he gets on staff at a hospital so he can have time with his family, something I never did with him. He says he doesn't hold it against me, but I know without a father and a mother who was always working, he had a lonely childhood."

"Does he know who his father is?" Charlie asked.

Elizabeth paused for moment before answering, "Yes, as I told you, he is more than my son, he is my friend. Friends don't keep secrets. They somehow eventually come out and then they can cause problems. Robert, that's my son's name, well, he knows about Bill and the time we spent together. He also clearly understands that Bill had no idea about his birth and I was the one who wanted to keep it that way for years. I also told him when he found out, he wanted to do the right thing by me and Robert. By the way, Robert was my father's name. By the time I told Bill of Robert's birth, I had secured a research position and was financially well off. He was then married and had his own family. Did you know Bill told his wife about our son? They even met one time. No, there is no problems for me or Bill when it comes to the decision I made long ago. I wish now that I had told him of our son sooner, but that's water long under the bridge"

Charlie thought for a moment before responding. "I guess in some ways that's why he asked me to help if I could. If somehow it was discovered that his father is a sitting federal judge, some news reporter could try and make something out of it. Not relevant to the case, but a distraction for your defense. He wants you to get a fair trial with no distractions."

Charlie then changed the subject, "Does your son know about the termination and the charges against you?"

"Yes, I told him right away. I rather he heard it from me, than seeing it on some damned social media site."

About then, a group of well-dressed men entered the dining room. Among them was a tall distinguished man who seemed to be the center of the others attention. They were laughing and seemed to be praising the tall man. In a moment the waiter escorted them to a private dining area just off the main room.

When their waiter returned, Charlie asked, "Who was the celebrity that just came in?"

The waiter took on a bit of a surprised look, "Sir that is the Reverend Jedidiah Cain. Do you not know of his great ministry to God? Why, almost anywhere in the city you can see his temple of faith upon the hill. At night it is lit so God himself can see the glory of Jedidiah's Church. And that's not all, each summer he holds a great prayer revival on top of Spruce Knob. Thousands of the faithful attend to hear him speak the words of God. He tells the people from the highest place in the state that he's brought them there to be above the sins and evil of the world below. It is a grand show. I have attended many times."

Charlie looked at the waiter as he said, "It sounds to me like you are a true believer of this man. Does he really speak for God?"

"Speak for Him, no mister, God speaks through him to those who truly believe. Someday God will return and Jedidiah will stand with him."

Elizabeth and Charlie looked at each other with a bit of disbelief as to the man strong beliefs.

Finally Elizabeth said, "I am sure he is a good man." As she smiled at him the man made the sign of the cross on his chest and said, "God loves him and he loves the true believers like me."

At that point Charlie was ready to have a cup of coffee and watch the sun set.

In a few minutes they had made their way to the outside deck. As Elizabeth sat down, the last of the evening sun shone on her. Charlie said as he sat down, "I'm sorry, but it is just hard for me to understand that people can be so gullible to believe a man in a $5,000 three-piece suit is their Messiah."

He looked at Elizabeth and asked, "Have you heard of this man?"

Elizabeth shook her head indicating she had not. "Truth is, Charlie, I have never taken the time to believe in much of anything, especially not what that man is selling. I have seen that cathedral lit up at night on the distant hill top. A great white cross a blaze with spotlights. You can see it for miles. I just thought it was some sort of monument, you know, kind of like those three crosses you see alongside roads in different locations."

About then her phone rang. It was Downing. Her face took on a pained look. After the phone call ended Charlie asked, "Is everything okay?"

She looked at him and said, "The grand jury indictment was just handed down. I have been charged with felony embezzlement, wire fraud and misappropriations of university research funds, what's next?"

Charlie tried to give her a look of comfort, "Well, I am not your lawyer, but I believe he will have you turn yourself into the court and have bail set. The process has begun and it will take a while."

She wiped a tear from the corner of her eye. "I just can't believe this is happening to me. Yes, I knew it could, but I just somehow never thought it would really come this far. I figured it was just a mistake and the government would look into the accounting and find I have done nothing wrong with the money. All I have ever worked for is to find answers locked inside our genetic code."

Charlie sat back and sipped his coffee, "Elizabeth, remember this is only an indictment. You are charged with a felony and it was brought before a grand jury. It is just the government's side of the truth and the particular part of the law they content you violated. Your side will be the reality of the truth in what they are contending. A jury will have to decide between the two.

Chapter 9
The Reality of Truth

Charlie tried to give her a look of comfort, "Well, I am not your lawyer, but I believe he will have you turn yourself into the court and have bail set. The process has begun and it will take a while before a court date is set. I would suggest your attorney make a motion for dismissal. Do you have the money or property that could be put up for collateral to satisfy the bail company?"

She looked lost as she thought for a moment. "How much money we talking about and what do you mean a motion for dismissal?"

Charlie took a drink and sat the cup down slowly, "As far as the bail goes, hard to tell, depends on the prosecutor in some ways. If he sees you as a flight risk, he might ask for a very high bail being set. And it depends on the money they think is involved. If they can't find some of the money and they think you could return to one of the countries you visited in the last couple of years, it could be very high. Your lawyer has to convince the judge you are not going anywhere. You'll need to take your passport with you and have your lawyer offer to surrender it before the prosecution gets a chance to make it an issue with the court. Judges like things tied up in a neat package. Shows your intent to cooperate with the court. As to the motion for dismissal, your lawyer needs to find some reason to petition the court to dismiss the case. In a sense it is saying the case is without merit and should not take up the courts time. The prosecution has to respond to the motion. Sometimes you can figure out what the prosecution is thinking if the judge gives any indication he is willing to listen to the motion. Your lawyer should submit a pre-trial brief explaining why he feels the case is lacking credibility and evidence of a crime, especially a felony. It is a bit of pre-trial wrangling that gives him time to look at the prosecution's case. The presiding judge won't want any lose ends. If he allows the motion in and the prosecution does not challenge it, he realizes you could be using his ruling to ask later for mistrial or an appeal. An appeals court will only look at the evidence, testimony and the motion submitted at trial. No new evidence or testimony may be entered, just a ruling. Was legal procedure and the letter of the law followed?"

She wiped a tear from the corner of her eye. "I just can't believe this is happening to me. Yes, I knew it could, but I just somehow never thought it would really come this far. I figured it was just a mistake and the government would look into the accounting and find I have done nothing wrong with the money. All I have ever been interested in is my work."

Charlie sat back and took a sip of his coffee, "Elizabeth, remember this is only an indictment. It is just the government's side of the truth and the particular part of the law they contend you violated. Your side will be the reality of the truth in what they are contending you did wrong. So far the grand jury has listened only to the government's side and handed down the indictment. They have not heard one word from your side of the story. In a court of law the government has to make its case to the jury and your lawyer can dispute and rebut their contention of guilt. After all that, a jury will have to decide where the two meet and if you are guilty of deliberately breaking the law or you are innocent. They will judge you on the letter of the law the prosecutor tells them you violated. You have to remember, this is about the money, not necessarily the science or your intent for research. The government prosecutor in a one sided argument before a grand jury has decided you broke the trust of the grant proposal and charged you. Embezzlement cases are difficult for the average juror to understand. If the prosecution makes it too difficult to understand with numbers and ledgers, jurors become confused and may not be willing to send someone to prison for what they don't really understand. On the other hand, if they make it simple, it allows your attorney to pick it apart. From what you are telling me, you don't think this is about misappropriations of money, but rather did your research overstep the grant guidelines. To be honest, I am not sure even I could make that determination at this point."

Elizabeth sat up in her chair as she thought for a moment before responding, "Charlie, my case has to be about more than just the money. It has to be about the research and what we have done to further science and medicine. And even bigger than that, what we have done for mankind."

Charlie shook his head, "If your lawyer is going to go down that path, you'll have to get the judge to let your attorney open that door to new evidence. And if the prosecutor doesn't want you muddling up the waters of his embezzlement case, he may object to your side getting that line of questioning into evidence. But, why are you so sure the research will make a difference? The question is, did you spend monies on anything other than the grant study?"

Elizabeth took out a piece of paper and began writing down some of the questions Charlie was raising. She looked up at Charlie, "You know that we are talking about a grant proposal that was originally written nearly 30 years ago. I don't mean the same one, but over the years we added to and tweaked it to keep up with the changes in the genetic science. When we began nearly 40 years ago, much of what has become reality today was only theory. Today we have moved beyond the theory of the science into practical applications. Charlie, do you know today we can harvest stem cells from a newborn baby's umbilical cord. Then it is frozen to preserve the child's stem cells for the future. We know those cells contain the blue print for the child at birth. In the near future a stem cell can be tricked into being a pancreas cell. If that child develops childhood diabetes, a new undamaged cell can be given in gene therapy and begin to regenerate the pancreas cells. They are doing clinical studies today and the results are very promising. You have to understand, when I first began getting funding for my research, this type of science was not possible. So, how can they judge me on using the funds inappropriately when the field of study has changed?"

Charlie shook his head that he did not know. "I know it seems messed up, but when we get into the court, the why and how will come out." He leaned toward Elizabeth as he asked, "If you know why this came about, when you talk with your lawyer you have to be honest. You are closer to this than anyone else. You know what your work is and where it's going, don't let the jury hear about anything from the prosecutor first, your lawyer can put the best spin on anything you are into that might have triggered this. Don't hide anything from your lawyer. And one more thing, I don't suppose you have that original grant application and the ones that followed?"

Elizabeth said, "I am not sure. I know there are a lot of papers from those early days in a file cabinet in my office, but I have not looked at them for years. There may be something there. Do you think the application could help?"

Charlie shook his head indicating he was not sure. They both looked towards the lake as the last shimmers of sunlight faded from its surface. Charlie knew there was more she was not telling. He had seen people with secrets before. He just wasn't sure if she was hiding something or she had not paid attention to details of her program over the years. But, so far she had offered nothing to answer his questions. One other thing, she had not said anything about the computer files and cryogenic canister Clay said was missing.

Finally she said, "Are you going to be working with my lawyer?"

Charlie thought for a minute before answering. "Got to be honest with you, I think you are hiding something and I don't like surprises, especially in the courtroom. You think about what I have said and we'll talk about it tomorrow at your lawyer's office. Don't forget, he has to want me to be part of your defense team."

Elizabeth laid her hand on Charlie's hand, "I have told you the truth, but the reality of the truth is not easy to find, even for me. My research has done wondrous things in my reality of science. I am not sure you and others can understand those ideals."

Charlie looked at Elizabeth and said in the tone of a prosecutor, "What is the ADAM project?"

Elizabeth did not looked surprised at Charlie's question. She stared at him as if waiting for more from him. Finally she asked, "Do you believe in God? I mean do you really believe that we began with God and his explanation of our beginnings in the book of Genesis?"

"Elizabeth, what I believe or don't, has nothing to do with your being guilty or innocent. I am not totally ignorant about what some say about mankind. If you're wanting to know, do I believe, *God moved his hand across* the land and man was there or do I believe we crawled out of some primordial soup as a slimy amoeba with a brain? I guess the truth is I don't believe either. Because if I did, it would mean I have a scientific belief in the amoeba theory or a type of faith in the Bible. I am just Charlie Morgan. Nothing godly about me or a green amoeba either. If there was a God, he let my family die for no good reason. He let fate kill them. That is saying they didn't have any luck that day. I am sure in the Bible it would tell me God has a plan for us all. That belief makes no sense nor reason to me, it just happened. A drunk went left of center and killed my family. They are dead and gone. No hereafter… no reincarnation, just the fact we all live and then we die. End of story. What are you going to tell me next, your research can bring back my family like that damned mammoth elephant in Korea?"

Elizabeth got a defensive look on her face as she blurted out, "NO! I am not going to tell you anything like that. Genetic science is not about bringing the dead individual back to life. It is about improving life at the most basic level for the living."

Charlie became somewhat defensive in his next comment, "You know, some would say what they are doing with the dead mammoth DNA is like bringing back a Frankenstein elephant. A piece of DNA from here and apiece from there, and one day you have a great hairy Frankenstein Mammoth. How much are they going to charge people to see one of those in a freak show? Is that what your science is trying to do?"

Elizabeth stood up, "Mr. Morgan, I think we both had better end our conversation for tonight. I would ask you to be at the attorney's office at ten in the morning if you think you can set aside your preconceived knowledge about my research? If anything, our conversation has shown me that with just a little twist, my case could be blown all out of proportion and my life's work would be just a trophy on some prosecutor's shelf. I just hope if you plan on working with my lawyer, you believe in something besides yourself and your self-pity. My science is neither good nor bad, it is science to improve mankind's future. If you can accept that, I welcome your help, if you can't, maybe you should go back to fishing and feeling sorry for yourself."

Charlie realized the conversation had gone too far, but he also learned something else. She was passionate about her work and he was sure there was more to her story than she had revealed.

"Elizabeth, I am sorry I expressed myself in such a manner. I may have had one too many whiskeys. I will be at your lawyer's office at ten in the morning. We'll see if he wants my help before you turn yourself in."

The two shook hands and said good night. Both firm in their position and both realizing there was something the other was not saying.

But, as Charlie was walking away he turned back toward Elizabeth as she crossed the deck. "Do you, Elizabeth Mathenson believe in God?"

Elizabeth stopped and for a moment and stood, not answering or even looking back at Charlie. Then she slowly turned to answer his question, when a voice from a nearby shadow said, "My goodness… in the middle of this tranquil setting I find two lost souls asking the very question God put me on earth to answer."

Both Elizabeth and Charlie looked toward where the voice came from. A puff of smoke gave away the source of the statement. It was a man, he was smoking one of those new electronic cigarettes. He was leaning against the door frame just off the main porch. After a moment they recognized that man as the preacher their waiter had told them about earlier that evening.

"Excuse me for overhearing your question, I didn't mean to eavesdrop, I just slipped off from my dinner party to satisfy my need for a smoke. I have been trying to ween myself from the real thing by using these new electronic ones. A terrible habit I picked up as a kid. My friends picked up the need for drugs that were easier to find than my cigarettes were in a small coal town in the southern part of the state. I remember the local cops would walk right by a kid with a joint cupped in their hand to bust my ass for smoking. Don't know if it was the cigarette or because my dad was head of the union in a small coal town. Either way, I got hooked and have been working to kick the habit. I ask God every day to stop this addiction."

He stood away from his leaning position and offer his hand to Charlie, "Forgive my manners, my name is Jedidiah Cain."

Charlie cautiously took the man's hand. Jedidiah placed the electronic device in a plastic sleeve and slipped it into the inside jacket pocket of his very expensive suit. As he turn to Elizabeth. "Good evening ma'am. I would like to know the answer to the question this gentleman has asked you. Do you believe in God?"

Elizabeth looked Jedidiah in the eyes and responded with her own question, "Do you have the answer? If you are the man with the glowing cross on the hill, they say you have the answers, and if you don't, God will speak through you and tell the rest of us the truth."

"Ah, a lady with a preconceived idea of my ministry. You see the cross I have erected to his glory as arrogant and boastful about my religion. I see that magnificent cross as a guiding light to those souls lost in the darkness of our world. I do not go into the streets and drag people into my church, they come freely to find answers. I simply offer the words of the Lord to those in a world tainted by corruption and greed. I am merely a messenger to tell those who will listen, he is coming and you must prepare yourself for that day. If you want your soul to live forever you must find the grace of God."

Charlie chuckled at the man's words. "Yea, that is about how I figured it to be. The only thing man has ever really known, he is not going to live forever. But, some religious person comes upon the idea, if they can convince these people that if they only believe in the words I tell them, they will live forever. Those words promise eternal life. That promise became the basis for the church and its power over people. As long as the people wanted to believe in eternal life, then the only place they could find that life, the church, would be the power on earth. Kind of like your friends growing up, drug dealers got them hooked on the need for another fix and the only place they could get their fix was return to the dealer. In your case, you tell them you have the words and the only place they can redeem them for salvation is in your church. Isn't that about how it happens?"

Jedidiah's chest seemed to fill with air as he placed his hand on both Charlie and Elizabeth's shoulders as he spoke. "My children, God is the light, he could enter each of your minds and show you the truth. But, he is a God that believes you must find it on your own. I am merely part of that path to self-belief and truth. I am not going to force you to believe, but I would invite each of you to join my congregation and hear the words. It costs you nothing."

He paused for just a moment and then leaned toward them as he said, "Or…it may cost you everything."

Charlie said in a skeptical tone, "If it costs nothing, how do you afford to build that cross? And I am told you have a great cathedral of a church. I know God doesn't sign the checks for that kind of stuff."

My brethren, "God merely opens his heart and mind to believers, those who are true believers know the word of my ministry must be spread across the land. And in the world of man, it takes money. They freely give and I ask for them to give even more. Money serves no man but through his faith."

Charlie laughed, "Whatever happened with that old saying, a rich man trying to get into heaven is like trying to push a camel's ass through the eye of a needle."

Jedidiah looked at Charlie, "Brother, I think you should brush up on your quotes of the scriptures, that's not how it works or said. I believe you have no faith in God. I think you even have less faith in man."

About then a voice from the doorway called to Jedidiah. "Mr. Cain, your party is waiting for you to speak to them before we part for the evening. Could you please return to our gathering?

Don't forget you have some important people waiting for your words."

Jedidiah motioned for the man to return to the gathering. "I am sorry, but I must return to my duties. God's work is never done for the penitent. I hope you two think about your beliefs and someday join my congregation." With those words he started to turn and stopped. As he did, he reached into his pocket. He handed each of them what appeared to be a business card.

"If you ever want to attend our church, the address is there. And if you ever feel the need for guidance, just move the card from side to side to see the words of our Lord."

With that he turned and followed the man toward a back room. Charlie looked at the card. He turned it slightly left and then right. He first saw a picture of Jedidiah with the information about his church. As he tilted it to the opposite side, a picture of God appeared, with the words. "God Loves You. Come join us in the Ministry of Jedidiah"

Charlie shook his head, "I'll be damned. I haven't seen one of these 3-D cards since I found them in a bubblegum pack when I was a kid. When I move that card, it showed a pitcher throwing a ball. I thought that was pretty special back then, but this one doesn't do much for me. Probably not even a real picture of God."

Elizabeth paid little attention to Charlie's comment, except to say, "Doesn't that image of God look a little like our new friend Jedidiah?"

They both laughed at her comment. She then looked at Charlie and said, "I am sorry we ended our conversation on a sour note earlier, I would like very much for you to be part of my defense team. And I'll tell you the same thing as I told Downing, I don't have a lot of money, but I'll pay whatever it takes to end this nightmare about my research. And, as to your question, do I believe in God? I haven't often thought about it. I do know that the wonders of life are a very complex question. I have come to believe that God and science sometimes are not so different. Both take a little faith to believe they can work. Without faith and the quest to ask why, man would still be living in the forest eating nuts and fruit. Doesn't the Bible say Eve offered Adam forbidden fruit? After they ate, their eyes were open to good and evil. I believe that good and evil was the awakening of knowledge and it was the key to man finding his own way in the world. He was given the opportunity to pursue paths in life that may not always be good. He may find that quest for knowledge can bring benefits to the future of man. But, that same quest could bring man to search for the genetic origins of himself. What if that search tells us something the Bible cannot. Can't that belief to search for the most basic questions be alive in each of us, as we pursue the very essence of mankind? My research did not start out looking for those answers. But, over time, along with other scientific breakthroughs, doors were opened just far enough, I and others decided to look inside the atoms of mankind. Could that knowledge given to all of us by Eve's actions be the same that gives me the ability to ask, 'What makes man who he is?' You asked about the ADAM Project. It is the key to my ultimate goals, to make our life better, but first we must understand what makes us what and who we are. I choose to believe that we are only doing what God intended, when he placed the tree of knowledge in the Garden of Eden. In his wisdom he could have stopped them from eating the fruit or even removed the tree entirely. But, he didn't. He chose to let us find our own way. Even if the final answer is the creation of man. Think about it, he did for us what we do for our children. We tell them the right way, but in the end they must make those choices themselves.

We can just hope they make the right ones. God simply did the same thing for us as mankind."

Charlie looked at Elizabeth as he said, "You look at the world and ask why, I look at the world and say, 'Show me proof.' I need that proof to keep you from going to jail. Simple as that. This can't be about all that scientific and theological stuff. God and his disciples are not going to be in the jury box."

Charlie looked at his watch, "I think it is getting late and I have had a long day. I'll see you in the morning. Remember, this is all about one simple question, like someone said in a movie, 'Show me the money!'

Elizabeth got an odd look on her face as she tried to understand his phrase.

Charlie realized she did not understand his reference to the line from the movie. "You don't get out much, do you Elizabeth?"

Elizabeth paused for a moment and then looked at Charlie with an inquiring expression. "I asked earlier… if while during your investigation of a legal case, have you ever discovered information that not only changed how you addressed the case, but changed something in your personal life?"

Charlie thought for a moment before answering. "No, I can't say that I have. A case is about evidence and the direction it takes me. It affects others, but not me personally. No, I can't say I have ever learned anything that changed my life. The only thing that ever changed my life was the loss of my family. What about you Elizabeth, why ask me that question…is it important to you that I think about your question?"

Elizabeth shook her head indicating it was not important as she said, "It is late Charlie, and we have a long day tomorrow, if I am understanding your conversation this evening."

He shook his head in agreement as they left the dining room.

Chapter 10
Stonemore

Charlie was up before the sun had shown it first golden rays on the tranquil lake outside his window. There on the lake he could see two Canadian geese moving slowly over the waters of the lake. Trailing behind were a dozen newly hatched youngsters trying to keep up with their parents. He wondered how the adults swam, showing no evidence they were moving their webbed feet below their bodies. The young geese had not yet mastered the art of swimming with no effort while keeping up. For a few moments he was lost in the serene image on the lake.

In a moment a knock on the door disturbed his watching the geese. Charlie wondered who could be bothering to knock on his door so early in the morning. He peered through the peep hole and saw a man in a sports jacket with the hotel emblem on the pocket. He opened the door.

"Sir, are you Charles Morgan?"

Charlie answered the young man, "Yes, that's me, how can I help you?"

"Sir, we have a phone call for you at the front desk, the person didn't want us to put it through to your room. He said it is urgent that he speaks with you. We tried to reason with him, but he insisted on talking with you in person and only at the front desk phone."

Charlie thought for a moment and realized it must be important, but he could not figure who would want him so urgently and under so strange a circumstance. He had no family, so it could not be a family crisis. After a moment of thought he said to the man, "Yes, please give me a moment to dress and I'll be right down."

The man politely thanked him and apologized once more for disturbing him so early in the morning.

Charlie wasted no time pulling on his pants and buttoning his shirt. He quickly looked in the mirror and ran his fingers through his hair to give it some semblance of being combed. He stepped into the hallway and began wondering who may be wanting him on the phone. He couldn't think of anyone that needed to contact him this early. As he entered the lobby the man who came to his room along with a young lady stood looking at him with an inquisitive look. As he approached the desk the manager pointed to a phone that sat on a small table in a sitting area near the front doors.

"Sir, you can take your call on that phone over there."

Charlie took a deep breath before picking up the phone and answering. "This is Charlie Morgan, who is this?"

On the other end, a voice sounded hushed and hurried, "Charlie, it's me, Jake, you need to listen to me… go right now and buy a burner phone. Don't ask why, just do it. I will call back to this phone at nine o'clock. Answer and give me the number of your new phone. Do you understand?"

"But why Jake, just tell me what you want to say. What is all this cloak and dagger stuff? And burner phones, what the hell?'

"I'll explain at nine, just do it now." The phone on the other end went dead and there was only the hum of an empty line.

Charlie sat for a minute as he thought about this strange behavior from his friend. He knew Jake was paranoid, but this just did not make sense, even for Jake. But, he knew if Jake was spooked about something it was for real, at least in his mind. Charlie stood up and returned to the desk and smiled as he said to the two curious people behind the desk. "Nothing to worry about, my neighbor forgot and let my dog get out. He's afraid I'll be angry with him."

The desk workers smiled and added some small talk about how they understood and made reference to a similar situation in their own lives.

Charlie started to walk away and then stopped, and turned toward the two and held up his cell phone, "Where can I get minutes for my phone? I must have forgotten to renew them and that's why my friend could not get a hold of me."

The man who came to his door pointed out the front entrance and told Charlie, "Down the road to the left. There is a quick stop station that sells minutes, they can help you."

Charlie returned to his room and finished getting ready to go out. He looked at his watch, fourteen minutes until eight. He knew he had just over an hour to find a phone and get it set up. Still he could not figure why Jake wanted him to get a burner phone and why the urgency in his call.

It did not take long for Charlie to find the quick shop and purchase a phone. The young man behind the counter offered to help him set it up. The boy had obviously done it many times before. He entered the information activating the phone. Charlie answered a couple questions to complete the service. By eight-thirty the phone was working and Charlie had written the phones number on a piece of paper and placed it in his pocket.

He stopped by a fast food business and picked up a cup of coffee and a breakfast sandwich. Before long he returned to the lobby and sat down near the phone. The two people behind the desk looked at him and asked. "Are you expecting another call?"

Charlie shook his head indicating, yes, "My friend doesn't know my phone is now working again and he said he would let me know at nine if he found my dog."

The clerks looked at the clock on the wall. It showed three minutes until nine. With one minute to go the desk phone rang, the man answered and pointed towards Charlie, "Just one moment, he is here waiting on news about his lost dog."

The phone rang beside Charlie, he picked it up and answered. The voice on the other end said, "What's this shit about your dog?"

Charlie responded, "Never mind, here is the number."

He fumbled with the paper from his pocket as he unfolded it. "304-266, he whispered the last four numbers to Jake so the desk workers could not hear.

The line went dead as Charlie stood and smiled at the two nosy clerks behind the desk, "Good news, he found my dog."

The two smiled and said, "Thank God. Our dogs become like our family. I'd hate to hear you lost yours."

Charlie waved as he stepped outside to sit in his car. In a minute the phone rang. "Hello, Charlie is that you?"

"Yes, Jake it's me, what the hell is all this burner phone and hush hush shit. Why can't you just tell me what you want?"

"Charlie, the number I am calling you from is a burner phone I purchased last night. You and I are the only ones that knows these two numbers, let's keep it that way."

Charlie was becoming a bit frustrated by his tone, "Jake what's going on?"

"You want to know what's going on, well I'll tell you. After you left yesterday, I got a visit from two FBI agents. You know the kind of FBI guys with sunglasses and no sense of humor. They also had a third man with them. He never said anything to me. But, he did ask one of the agents a question about my background. The guy had a foreign accent. If I didn't know better he sounded Israeli. Anyway, they wanted to know in no uncertain terms why I hacked the university's main server. They explained that I caused the site to crash. They knew after the system rebooted that I made my way into the Genetics Departments web pages. Then they pointed out the genetics department and some of its personnel are under Justice

Department investigation. Then they said my intrusion could be considered as tampering with evidence in a federal investigation."

Charlie listened and then asked, "What did you tell them and how did they track it back to you so quickly? I thought you had some fancy firewall to prevent being back tracked or invaded with a worm or virus tracker program?"

Jake said, "I told them some cock-and-bull story about my working with the senior's home and them wanting to know if genetic science could help with Alzheimer's. One of the old ladies there said her grandson did an internship while he was studying there. Supposedly he told her they were doing a lot of research to find new cures for the diseases. I don't think the agents bought it, but they did not ask any more about the university. The older of the two said he remembered me when I worked for the justice department as a forensic accountant. The other agent asked if I was still doing consulting work for the department. I told him not sense my wife died and I retired to catch fish and help at the retirement home. I am guessing they were willing to take my word for my activity… on this occasion. I'll not get away with it next time. In fact if there is another visit from the FBI I will unceremoniously be loaded into a black sedan and taken to the field office for questions and not on my front porch. I'll not make the mistake of being so careless with my next internet inquiry. I'll take more precautions, just didn't think it was that big a deal about this case and my firewall should have kept them out. Obviously I was wrong. This thing you got me into is like some damned Atlantic iceberg, a little showing on the surface and a hell of a lot below. My looking into the university department should not have triggered this kind of response from the FBI at least for the information and reason that you gave me concerning the doctor's case. We never even got close to finding the path for the money. But, after this

I suspect this is about much more than money. Did you find anything out when you talked to the doctor?"

Charlie did not answer for a moment as he thought about what Jake had told him. Then he asked, "Is that why you wanted me to purchase one of those phones which you buy minutes to use? What did you call it… a burner phone?"

Jake quickly responded. "Yea, that visit from those guys in the dark sunglasses and one with a heavy foreign accent sure got my paranoia going good. If they are playing for real, they already have my phone tapped to find out if I contacted anyone. All they have is I called the hotel. I am guessing if they are real serious about this, they will visit the hotel this morning sometime and ask about the phone call and who received it. They'll track it to you pretty quickly, but I'm hopin' they'll not figure two old government workers are smart enough to get burner phones. Kinda' like were the bad guys, ain't it Charlie?"

"Jake we are not the bad guys. We are not secret agents. We are maybe going to represent a doctor who is about to be charged with fraud and embezzlement. You're letting your imagination run wild. If I get to work with Elizabeth's attorney on the case we'll get access to the evidence. No need to sneak around the internet. We'll have the right to look at the evidence. So you can put your secret decoder ring away. And I'll try and forget you made me go buy a phone with minutes. Hell, I already got one of my own. Never had much use for the ones with a contract tied to them. All that bullshit of apps, phone games and movies. I just want to talk on the phone and ignore it when I don't know the number."

Jake grumbled a few moments before asking, "Alright Charlie, what's next, sayin' you are right and this is just a money case?"

Charlie responded, "I am on my way into town to meet with Elizabeth and her attorney. I have to figure out my part, when working for this brand new lawyer. We'll see if he's willing to work with me or ignores me completely. Elizabeth may have to make a decision before too long. Jake, I'll call you when I know something on my end. Oh, one more thing, you said something about the guy at Stonemore who knew something about this type of research. Maybe he can help us understand what's behind the research and how this woman is involved in the overall picture. If you get anything, text me the word, "yes". I will get ahold of you when I have a chance. Now, you go back to catching fish off your dock. I'll talk to you later today when I know something. Maybe by then you'll have something from Stonemore." The two said good-bye.

Charlie wasted no time getting ready to head into town to meet with Elizabeth and her lawyer. He looked at the clock and it showed after nine already. He had very little time to shower and get dressed.

By nine thirty, he was on his way to town. He knew he would be a little late getting there, but he figured they would not start without him. He knew Elizabeth realized she was in so much trouble and would need all the help she could get. And she was no fool to put her life into the hands of a green lawyer.

Shortly after ten fifteen Charlie walked into the lawyer's office. Elizabeth and her attorney were sitting in the back office talking. "Good morning, do you have any coffee?" Charlie asked.

The young lawyer stood and offered Charlie his hand. "I am sorry about yesterday. I guess I got blindsided by your comments, wanting to help with this case. My first real case and you just walked in from the street and caught me off guard. And besides, I checked you out last night. I did remember you from school, but I did not know you were once a top prosecutor for the Justice department in these types of cases. But knowing that, I have to ask why? I mean, you spent your career on the other side in this type of case. And there are no big fees for our services in this case. For me, it is a chance to get a real case and federal courtroom experience. If I don't mess up and allow my client to go to jail, I may get a little recognition among my colleagues. For you, well, you gain nothing."

Charlie responded, "First off, you are right, I have nothing to gain. Secondly, you have to put out of your mind what you intend to get out of this case. Doctor Mathenson's career and freedom is on the line. For you it is a case, for her it is the rest of her life. You don't worry about yourself in cases like this. If you are looking at a case from how it benefits you, go into divorce law. Everybody there is already figuring they are losing something. The lawyer is the only winner when he gets paid."

By this point in the conversation, Charlie had found the coffee pot and poured himself a cup. He returned to the front desk and sat down on the corner of it. "For my part, let's just say I want to get back into the courtroom and leave it at that." He looked at Elizabeth, "Is that a good enough answer for you Doctor Mathenson?"

Elizabeth smiled and said, "I welcome anyone that gives up catching fish to help me. Now, I think we should talk about my case. I understand an indictment from a grand jury has been handed down. Does that mean a warrant has been issued for my arrest?"

Downing leaned against the door frame and said, "No, I don't think so, I called the U.S. Attorney's office and told them we would be waiting for a call as to when to appear and that I would vouch for your appearance."

He looked at Charlie, "Is that what I should have done?"

"Yes, that is good for now. Did they tell you who the prosecutor will be in this case?"

"No, they simply asked for my contact information. The docket clerk will call when she has a court time to appear."

Charlie looked at Elizabeth, "How much money and assets do you have clear title to? They'll need that information if the judge sets bail. You will have to make good on the securities to gain your freedom. Just how much money are they saying you took… misappropriated…. or whatever they call it, how much?"

Elizabeth got an odd look on her face, "I am not sure, depends on how far they go back to make the case. Over the years there has been millions of dollars from scientific grants, government program monies and people who had an interest in my research. For a lot of years no one paid too much attention. Then about ten years ago there were several major advances in cloning and mapping the human genome and our research took off. Money poured into research programs like mine. I didn't pay any attention to the program's finances when the big money started coming in. At least no more than I did when we struggled for funding. I just continued my work oblivious to the money. The department controllers were responsible for that end of the program. I was into the research and didn't pay any attention to the finances. My only real connection with the money was when the controllers were angry at me for not keeping them in the loop with my spending. That doesn't make things good for me…does it?"

Charlie shrugged his shoulders, "Depends. Whose name appears on the department as head of it? And if it is your name, did you sign off on any invoices or any papers concerning the funding at the budget meetings?"

She thought for a moment and then answered, "I guess I did. I probably signed thousands of papers over the years. And before you ask, no I did not take the time to read them all, I just signed. I am a doctor, not an accountant. Wouldn't the controllers have to be charged too? After all, they were watching the funding and writing grant proposals. They knew more about the money than anyone else in the program."

For a few minutes no one said anything and then Charlie said as he looked at Downing, "She is right, we will need to get a list of the controllers and have them checked out. I don't think they had anything to do with the question of the money or they would have been indicted at the same time. But, they may be able to shine some light on the money in the program and how it was typically spent. The prosecutor will look for money that deviated from normal spending patterns. If she spent a million dollars in travel and doing research out of the country before the_questions began, it is unlikely that travel or that same research pattern is a problem. Deviations from the normal every day is what they'll look at. They look at questionable lines of spending and follow the money. We will need the accounting records back at least fifteen years. Oh yea, one more thing. She said the original grant paperwork may still be in her files. If we can get a copy of them it may help to establish the guidelines of her research back then. We can compare it to a recent request. It may show she never had any specific guidelines in the direction of her work."

Charlie next looked at Elizabeth, "Do you know any reason why the State and Justice Departments would be interested in your research beyond the question of money? And another thing, do the Israeli's have any interest in your research?"

Elizabeth stood up and walked to the window and looked out onto the busy street. Then she turned to the two men, "FBI, CIA, who knows, in the last few years people from different government agencies held meetings with the science department. I never took the time to attend. I had no interest in those people and their investments in my work. A few times I also saw uniforms, you know, Army types. Lot of stars and stripes on their uniforms. Never asked anything, just looked.

The science department public relations person usually conducted the tours of the main labs."

Downing said, "You never told me about the government being at the lab. Why not?"

"It was never important to me. A couple of guys asked me about my trips to Korea and Israel. Wanted to know about my time there and what I was working on. They even asked if the Chinese or Russians were there at the same time as I was in Korea. I told them yes, the Chinese and Russian research teams were the ones digging up mammoth specimens in Siberia. I just figured they had the right to be part of the research."

James asked, "Why the trips to Israel?"

Charlie quickly injected into Downing's question. "Yea, Israel. I keep tripping over people from Israel in this mess. Any idea why?"

"Simple, I made several trips there. They are one of the biggest depositories of DNA and ancestry research outside of the Mormons. I was interested in some of their research and some specimens they had for sale. Just research stuff."

Charlie looked at Elizabeth and asked, "A minute ago you spoke of main labs, were there others at the university?"

Elizabeth took a deep breath as she thought about her answer. "Yes, there are labs where full security protocols are in place."

Downing hurriedly asked, "What happens in those labs?"

Elizabeth paused for a moment before answering, "I don't know how to answer that question. It is confidential research into programs not publicly known. I am bound by confidentiality rules not to divulge the information behind those doors. If they ask me that on the witness stand I can't answer. I know that would violate a bunch of rules and my career would be over, regardless of the trial's outcome."

Downing looked at Charlie, "You're the seasoned veteran of the federal courts. How do you handle this situation?"

Charlie responded, "I have only run into this with a couple cases. Both times it was government contractors and their funding was for top secret programs. Money was missing from the bottom line. We submitted questions to FBI investigators and they came back with some answers. Sometimes it helped and sometimes it did not. Every case is different. But, in those cases the questions were about tangible things, engine parts and weapons electronics. It did not require we see or know about the weapons systems, only whether the contractors were providing what parts or service that they agreed to. FBI investigators provide the collaborating evidence."

Elizabeth looked at the two men, "I can tell you this, the questions would have to be complicated even to be asked and require knowledge of biochemistry. And if you think the questions would be hard, the answers would be nearly impossible for you or a jury to understand without the scientific background to understand."

Charlie stepped toward Elizabeth, "Do you mean to tell me, it is impossible to understand what you do behind closed doors. It is important to the people who may sit in the jury box to have some understanding of your research. You aren't going to tell me that in some secret lab you and your colleagues are building a Woolly Mammoth?"

Elizabeth half laughed, "No, nothing as simple as that. Charlie you have no idea where science has taken us in the last few years. We are on the threshold of things that will change man's life for the next thousand years. We can look into human molecules and understand what we see. If we can see it and understand, we can find those things that cause diseases. In fact some gene therapy is already helping people with correcting genetically damaged cells."

Downing looked strangely at Elizabeth, "Ma'am you are giving me pause as to what you do over there, I remember hearing whispers when I did my research paper on genetics and the law. But, I guess in ways those things whispered about in the breakroom were stories somewhere between Star Trek and Frankenstein. Stories of scientist fooling with the very essence of man. Is that about right?"

Elizabeth sat for a moment before answering his question, "All I can tell you is, scientists all over the world have made wondrous discoveries. If some difficult questions and problems can be solved, it will revolutionize medicine as we know it. Cancer, diabetes, Alzheimer's and dozens of other diseases can be treated, if not cured, with these new sciences. There are those who believe this new science will alter man's future. For me, I am just working to find the way to help prevent human suffering far into the future. Someone else will decide if the science will be used, and how. In all the new science the government has an agenda for regulations. Drug companies have an agenda to make money. And fundamentalists from all religions have an agenda to keep God as the one true answer. I suspect this investigation was triggered by one of those factions wanting to bring light onto the program and me. I don't know that for sure, but I suspect that is what happened. Over the years, I and other members of our research team have received anonymous threats on our research, and even our lives. We knew the letters had to come from within some of the lab's own staff, someone who came to work here and on their own, decided to be judge and jury of the science. Some believe we should not be fooling with the very basic structures of life. There were others who believed it was in direct contradiction of the Biblical or moral compass of mankind. These types of feelings are not unusual in new research. When penicillin was discovered there were people who said it was supernatural poison because it came from a gray mold. They talked of a time when witches used mold brewed into magical potions to control people's minds for evil doings. There are some strange people out there still today."

Charlie asked, "Did you contact campus security about those threatening letters?"

"In the beginning, but I am not sure if we still did in the last few years. Like I said, just someone making idle threats. Some were personal, others thought they were messengers from God. Who knows for sure? Do you think they could have started this investigation into the department's finances?" She looked at both Charlie and Downing questioningly.

Charlie shook his head as he said, "At this point we don't know. We will have to file for disclosure of evidence and when we do we want to ask for any security files pertaining to the department going back…."

He paused suddenly and then asked Elizabeth, "Do you remember when the letters or notes began?"

She thought for a few moments, I'm guessing fifteen years ago. It was before the big interest in the genetics department began, and that was about ten years. Yea, I'd say fifteen years."

Downing scratched his head and then asked, "Can you remember if anyone in particular started in your department back then, I mean about fifteen years ago?"

Elizabeth shook her head no. "Over the years with new interns, lab staff and administrative people, it is hard to remember. But, I am guessing the personnel department would have the hiring files from back then. I just can't remember people that came and went or even stayed. Like I said, my main concern was on my work."

The three just sat for a few minutes thinking about the case. Charlie wondered if this was about the money or something else the lab was working on. And he had to figure how to see inside the department, and especially behind the security laboratories. Downing was writing down all the requests for information and how to separate the evidence from background information. Elizabeth was wondering who if anyone was still working in the labs.

About then the phone rang on the desk. For a moment Downing seemed to be waiting for someone to answer the phone. Charlie looked at him, "Don't you think you should answer?"

Downing snapped out of his thought process and answered the phone, "Downing Law Offices, how can I help you?"

Charlie could hear a woman's voice on the other end speaking to Downing. He could only understand a word here and there. He did hear the voice say more than once, "Wheeling." Downing seemed to only be able to say, "Uh-huh, yes, I understand, Uh-huh, I see."

Finally the phone call ended and Downing took a few moment to write notes on a pad. After he finished Charlie said "Well, was that about this case?"

Downing looked at Charlie and Elizabeth, "Yes, they want her to appear before Senior Judge Harold Thompson in the morning at ten in the Wheeling Federal building."

Downing looked up at Charlie and said, "I thought this would go to the Elkins court. Why are they sending us to Wheeling? We haven't reviewed the evidence yet or arrange to interview any of the controllers at the university. Is this normal for a trial of this type?"

Charlie had walked to the window and was watching cars pass by on the busy street. For a few moments he said nothing in response to Downing's question. Finally he turned and said, "No, it is not normal, but it is not unheard of either. We don't know what the prosecutor in the case is wanting at this point. But, it also gives us a chance to file a motion for dismissal and present the court a brief supporting our claim of lack of evidence. If we are lucky the judge will at least ask the prosecution questions before he rules. That may give us some insight into their case. Many years ago I had a case we filed against a company who got a research grant to create fuel from coal. Along the way they spent monies on side projects as the government and I saw their case. I will call my secretary and have her send over the file along with motions they filed along with their brief. Maybe there is something we can use in ours. I'll look it over and make adjustments to our motion for dismissal at this time. She'll fax it to your office. Make sure we have it for tomorrow's court date. Judge Thompson does not like motions and briefs that are dozens of pages long. He is a direct and to the point kind of a judge. We'll make our motion to the point and very brief. You need to remember that in the courtroom. He is not very tolerant of long meandering witness examinations. He'll tell you if you have a question or statement, get on with it. His court is not a stage for Shakespearean players. The federal prosecutor will already know that. The fact that their prosecutor doesn't know that I am helping will give us a bit of an advantage. But it won't last long."

Charlie looked out the window again thinking. Then he turned slowly to Downing and asked, "Did they say who the prosecutor is?"

Downing answered, "Yea, I wrote it down here on the paper. Harrison Dobbs." He looked at Charlie, "Do you know him?"

Charlie had sat down in the window seat of the front bay window. "Yes, I know Harrison. In fact, he got his start under me as an assistant prosecutor several years ago. Worked with me for three years, I think. He transferred to the Southern District when a prosecutor's job came open after a while. I didn't think he had returned to the Northern district. I wonder, did he return or did they bring him back for this trial."

Elizabeth asked, “Is he good?”

Charlie looked over his glasses at Elizabeth, “Yes, he is good, he was young and aggressive when I knew him. He has ambitions to be a federal judge someday. Hardliner conservative, doesn’t miss a trick. Pays attention to details. Good man, but even I felt like he was in competition with me on each case, even though I was the lead attorney for the prosecution.”

Downing looked at the paper and studied it as he said, “Why Wheeling? I know it is just a preliminary hearing, but why take it to Wheeling?”

Charlie glanced at Elizabeth as he answered, “Maybe the Elkins docket is full and the prosecutor wanted to get the pre-trial work out of the way. If Elkins is going to be tied up for a while, this makes sense. Maybe we will find out tomorrow.”

Charlie looked at his watch and quickly jumped up from the window seat, “I did not expect to get into this so quickly, I didn’t bring my suit or even my briefcase. I’ll have to head downtown and get a few things before tomorrow. I am sure Judge Thompson would not appreciate me coming into his court room in blue jeans and a red flannel shirt.”

Downing asked, “Do you know the judge?”

Charlie replied, “Yes, Thompson has been around for many years. I have argued several cases before him before we both retired. Likes things neat and tied up in a bow. He’s old school. In his trials, he likes to end the day’s proceedings about four. His clerk once told me he tends to drift off much past that. But, the hearing shouldn’t last long.”

Downing asked, “Did you say he was retired? How is he back on the bench to hear this case?”

Charlie picked up his coat as he answered Downing’s question. “If for some reason they need another judge in a district, it is not uncommon to bring back a retired judge. When they do that they’re called Senior Judges. I wouldn’t call this a special case. But for some reason he’s been pulled out of retirement and back to the bench for this hearing. I guess kind of like I have been brought out of retirement."

Charlie started toward the door and then stopped and turned around, "There is one very important question that has not been asked of Elizabeth. Are you guilty? Tomorrow is about the grand jury indictments. After charges are read the judge will ask you to enter a plea, guilty or not guilty. Our job tomorrow is to make a motion for dismissal of all charges. File a short brief to support our motion. If the judge rules against us which I expect, he'll ask for your plea. Harrison will want to look aggressive and ask for no bail in your case. We'll argue otherwise. So Elizabeth, what is your plea, are you guilty or not?"

Elizabeth looked at both of them, "Not guilty Mr. Morgan" was her reply.

Charlie smiled and said, "See you here at six, Wheeling is a two hour drive. We can talk on the way there and still give us a couple of hours at the Federal building. I like to see what we can pick up and get a feel for the lay of the land before we go before the judge. If we see who comes along with the prosecution, it will give us some idea of how serious the game is. Elizabeth, when we get into the building and the courtroom, I want you to look around, and see who's there that you may know. I want to know who all the players are. Right now, we have no face cards showing in our hand, time to ante up in the pot and ask for some cards. Maybe push a little bluff if we need too. Strong confidence and two attorneys representing her may shake their strategy when they go before the judge tomorrow. They know about you, James, but they may not yet know about me. I suspect it will get them to question what we're up to. I'm betting they figure just you and Elizabeth will be coming into the courtroom tomorrow. A young green lawyer, no offense intended, and the defendant who is only interested in her research. Yea, I am counting on my being there to raise questions and a few eyebrows. I'll see you all at six."

As Charlie walked to his truck, he felt the cell phone vibrate in his jacket pocket. He removed it and looked at the small screen. It indicated he had a message. It was simple, "Yes and no."

Chapter 11
The Stonemore Question

As he was about to decide what to do with the text message, he got a call on his old cell phone, it was from Nancy in his office. "Charlie is that you?"

He answered, "Yes, did you get my message about the coal company research project file?"

Charlie could hear papers rattling as Nancy was looking through the information. "Ah, well yes and no. I could not find the judge's response on the motion for dismissal. But, I did find the brief and the pre-trial motion filed by their attorneys. What do you need with this old information anyhow?"

Charlie paused for a moment before answering. And when he did…he did not answer her question, he only gave her instructions. "Nancy, fax those documents to my hotel. Call them ahead of time and tell them to add the cost to my bill. Then tell them to slide them under my door. When I get in tonight, I'll review them. I'll prepare our motion and brief and fax the corrections back to you. I need them typed and returned by morning. Email or faxed them to James Downing's office in Monongahela City. I'm sorry but this will be a long night for you, but I haven't any other options on short notice. You have written enough of these that you know where to add in the legal stuff. I'll just hit the high points and you fill in the rest. Thompson is the presiding judge, you know how he likes short and to the point briefs. I'll make sure there is a little something extra in your Christmas stocking this year."

Nancy grumbled a little before saying, "Charlie Morgan, I thought you retired. If I really believed you have, I'd up and retire on you. But, I guess you still need me and I need you." She chuckled.

Charlie said, "You're a good woman Nancy. I'll be looking for those documents shortly. Call you if I need anything else." The two said good-bye.

Charlie's attention returned to the burner phone as he tried to decide what Jake's cryptic message meant. Then he remembered. He had instructed Jake to send the one word message if he learned anything useful from his Stonemore connection. But why had Jake texted both yes and no. He decided the unclear message would have to wait until a little later. He had a suit to buy and a briefcase to find. As he approached his truck he looked at his reflection in the window glass. He thought to himself, you could use a haircut and good shave before going into court tomorrow. Since retiring, he had not gotten a routine haircut. Nor had he shaved before leaving the house each morning. Fish didn't care and his office secretary would complain no matter how he looked, even if he wore a three piece suit to the office. She once told him he looked, "Frumpy". He was not sure what that meant, but he figured it meant he was thirty pounds overweight, wore his hair too long, his suits were a little on the large size and he wore suspenders. His appearance may be "frumpy" to her, but Charlie figured he was comfortable, that was all that mattered to him these days.

When Rose was alive, she laid fresh clothes out for him every day. She made sure his shirt was ironed and hanging on the back of the chair. Pants were pressed neatly and the pockets not full of junk. Charlie had the tendency to have a pocket full of coins, fingernail clippers, keys for his truck, house and even a set of his wife's car keys. All this tended to make his pants sag if she did not clean his pockets. He usually grumbled at her for annoying him about his clothes. Truth is, he missed her keeping him looking half decent. These days he settled for his secretary saying he looked "frumpy."

Charlie decided he would wait to answer Jake's text concerning the Stonemore question until later that evening. But, as he started his truck he thought for a moment. He decided he would ask Jake to join him, if he could, at the lodge and then travel with them to Wheeling tomorrow. Jake would know what questions he needed to ask Elizabeth regarding his request for information from the government.

Charlie pressed the call button and the only number in the phone appeared. He selected it and waited for the phone to ring on the other end.

On the third ring, Jake answered, "Did you get my text? Yes, I got a hold of my friend at Stonemore, and no he won't talk with me. Well, at least not over the internet or phone. In fact all he would say is, *"The steaks are excellent and the Irish whiskey is good at Parliament House."* It is a place we use to go have lunch near the university when I was doing research for the government a few times. Whatever he knows, he wants a face to face meeting to tell me."

Charlie paused for a moment before responding. He had not expected Jake to blurt out the answer so quickly when he called him.

"Well, I guess we'll see how this case goes. One of us may have to travel to Stonemore and talk with your friend. Jake, right now I need you to come to Monongahela City and go with us in the morning to attend court in Wheeling. You can talk with Elizabeth and make sure we ask for the proper information regarding the charges against her."

Jake paused, "Are you shiting me? I am not ready to go anywhere. You just said find out information from my friend in England, I tried. Now you want me to drop everything and join you for a road trip to Wheeling, is that right?"

Charlie responded quickly, "Yea, I guess that's about it. What do you say?"

Jake responded, "Sounds good. Gets me away from the cabin for a while. And you'll be buying me supper at the lodge, right?"

Charlie answered, "Yea, and I'll throw in a couple of beers."

Jake looked at his watch, "Charlie, it is two-thirty, take me an hour to get things ready and two more to get there. I'll see you around five-thirty."

Charlie said, "Okay, oh one more thing, can you do a background check. His name is Clay Gordon. He's a professor of law at the university. I think he is originally from Israel"

The phone was silent for a moment and then, "Clay Gordon, lawyer, university, got it. See you in a little while." The call ended.

As Charlie was heading toward the mall to find a suit and place to cut his hair, he wondered, why won't Jake's friend talk with him over the phone? And does he really want Jake to come to England for a few simple questions? He would find out more this evening. Charlie pulled into a parking space and took a deep breath. He disliked malls, or at least that is what he used to tell Rose when she went shopping. He would usually go by the book store and see if they had any new books on tying flies and then find a place to sit at the center of the mall. That location served two purposes; made it easy for his wife to find him and he could look at the latest batch of weird dressed college kids. He sometimes told his son Timmy strange whimsical stories about the weird looking kids and the fantasy world they descended from. He wondered if someday his son would be one of those weird dressed college kids at that mall. He would give anything to see them both one more time, even if Timmy was dressed in baggy pants with blue hair and a pierced nose. The reality of those thoughts reminded him how much he missed his wife and son.

Charlie wasted no time finding a dark suit at the JC Penney's store. He also found a place that took walk-in customers for haircuts. A young lady insisted on washing his hair before she could begin trimming it. Charlie knew she was just trying to make an extra buck by insisting on a shampoo, but it wouldn't hurt to have the twigs from low hanging tree limbs washed away. Truth was he didn't seem to mind a bit of pampering from the young lady. For her, it was a way to get a bigger tip. For Charlie, it was a few moments of a normal life. He remembered in the early days when Rose and he were first married, she would wash his hair in the shower. He had not thought of those days in a very long time. He realized it had been years since he had done things normal people do, go shopping and get a good sit down haircut from a girl with a tiger tattoo on her neck. In fact, it was just before he lost this wife and son. That terrible day became the defining line in Charlie's life. Before that Thursday evening his life was routine. But after opening the door and seeing the state policeman with his hat in his hands standing on the porch, everything changed for Charlie. He continued to work for a while, but whatever made his life special ended that evening. He had in some ways closed his memories to those days. The bright smile of Timmy when he caught the ball Charlie tossed to him in the front yard, or the feel of Rose's warm touch on his shoulder in the darkness as she said, 'Good night, I love you, Charlie Morgan." He usually responded, "Me too." He knew she realized he loved her also.

Charlie found his way back to his truck and placed the suit behind the seat. He looked across the parking lot and saw an office supply store. He remembered he needed a briefcase and writing pads and pens. He headed toward the store, trying to think if he needed anything else. He remembered, he would see if they had a pocket recorder for interviews he may need to do. He knew he had all this kind of stuff back in his office and if he had asked his secretary, Nancy, she would have brought the items to the lodge. But, somehow this gave him a sense of fresh and new as he was preparing this case. Something about this case seemed to intrigue him, he was just not sure what. For the first time in several years he felt an eagerness to return to the courtroom.

Charlie spent the afternoon shopping and had returned to the lodge. He spent most of the next hour writing notes for tomorrow's hearing. He tried to remember his conversation with Elizabeth.

Somewhere deep in his thought patterns he heard a knock on the door. It was followed by a muffled sound of a man's voice. "Charlie, are you in there?" It was Jake's voice.

"Yea, just a second. Charlie stood and paused for a moment as he made sure his knees were going the same way he was. He had sat too long without walking around and his lower legs had gone to sleep on him. After a moment the feeling returned and he moved toward the door.

He opened it to greet Jake who blurted out, "Well, I am hungry and you said you would buy my supper, let's go."

Charlie shook his head as he said, "Jesus Jake, at least give me time to put on my shoes. This place is a stickler for people wearing shoes into the dining room."

Before long they were being escorted to their table. The waiter seemed to sense they would be more comfortable away from the main dining room. He gave them a table near the back wall. "May I get you gentlemen something to drink while you decide what you're having?"

Jake said straight away, "Sure sonny, I'll take some kind of dark beer in a glass and if you see it is getting low, you fetch me another right quick."

The young man smiled at Jake's phrasing of words and loud tone of voice. He then looked at Charlie as he said, "Yes, I'll have a scotch on the rocks and an iced tea with my meal." The waiter thanked them and excused himself.

"Well Charlie, what did you find out today and why am I going to Wheeling tomorrow?"

Charlie answered, "The indictment was handed down and instead of taking her to Elkins, they decided to send her to Wheeling. Tomorrow will be a preliminary hearing with bond and a trial date being set. We'll file a motion for dismissal, but I don't think it will be granted. I have agreed to help with Dr. Mathenson's defense. I would like for you to be part of her team. You were one of the best investigators back in the day. I'm guessing you still know how to find things out about people. Especially the things they don't want known."

About then the waiter returned and sat the two drinks on the table. "Have you gentleman decided on your meal?"

Jake looked at the man and said, "I'll take a steak, big, and cooked on the rare side. I mean make sure it ain't movin' and warm in the middle. You got that?" The man said, "Yes, I think I can remember that. Anything else besides your steak?"

"Baked potatoes and some kind of vegetable and hot rolls."

Charlie looked up at the man and said, "I'll take the same thing, except cook my steak to well. It makes me feel like I am a little further out of the jungle than my friend."

The man smiled a Charlie's comment and said, "Yes sir, I understand."

Jake took a drink of his beer and then said to Charlie, "I'm guessing this is a pro-bono case. You said nothing about money or compensation for my time and efforts."

Charlie took a sip of his drink, "To be honest, Jake, I never asked about money. This one I am doing because I need this. Not sure why, but there is just something about this case that sparks my interest. I know you probably don't have the same feelings, but I need your help. You are the best investigator I ever worked with. Didn't matter if it was in a hidden computer file or in the back room of an office building, you somehow always came up with the answers I needed. And I need you this time. I will pick up any expenses you may have, but we'll keep that between us."

Jake took another drink from his glass, "Charlie, I don't know if I ever told you this, but my wife's family was loaded. And she inherited lots of money. But, you know me, money was never very important. In fact, up until Ella died, she still would look in my wallet and see if I needed any. I just never really cared about it. I spent what was in my wallet and then if I needed more, I went to my credit cards. My work was chasing leads and finding ways into corporate accounts, that was a big thrill for me. It was like them kids with their warlord's games. Chasing the troll monsters through the different levels to win the game. For me, it was chasing the money through different levels and accounts. It was a game and I liked playing it. But back then it was legal, well for the most part it was. If I am going to help with this case, we need to be recognized by the courts and get authorization to find out what we can. Getting paid is not a problem for me. Getting back into the game is stimulating."

Charlie thought for a moment before answering, "Stimulating, well we'll see. This may just end up being a case of money spent on one type of research and the government wanted it spent on another. And if that is the case, someone, somewhere is driving this. There is something about this case that Mathenson is not telling me. She is holding back, I can tell from her eyes. Maybe our evidence discovery motion will help find some of the missing pieces. We'll see what tomorrow brings when we go before Judge Thompson."

About then the young waiter returned to the table and sat the service tray on a small stand next to it. As he picked up Jake's steak he made a sound. "Moo."

Jake looked at the man, "Smart ass, over cooking kills the flavor of the meat. You should try it sometime."

The young man picked up Charlie's steak and sat it on the table in front of him. "Sirs, is there anything else I can get you with your meals?"

Charlie looked up at the man, "Don't mind my friend, he is only used to catching crappies trapped in old computer cases and helping old ladies with their taxes. Doesn't get out into the refined world much."

Jake said in a quiet tone, "Moo my ass. It just better be good." The young man smiled and excused himself.

Charlie looked at Jake's steak and said as he cut it into bite size pieces. "Don't you know you can get worms from uncooked meat?"

Jake chuckled and said, "You can't get worms from slightly cooked beef, this is not pork. For a big time lawyer you don't know shit about your meats."

The two men ate and made only small talk for the next twenty minutes. The waiter returned regularly with a full glass of dark beer for Jake. Finally they finished and the young man was clearing the table and asked if they would like dessert.

Charlie spoke, "No, I don't believe I'll have any dessert." He then looked at Jake. He indicated he was not interested in dessert either. "But, we would like a cup of coffee."

As the man removed Jake's plate he said, "Was it to your liking sir?"

Jake said to the man who wore a slight grin as he asked. "Damned right I liked it, and you can tell my friend, a rare steak does not have worms or disease. It is just a good steak."

The young man removed the plates and silverware and said, "I'll be back with your coffee." As he turned to leave he looked back at Jake, "Sir, you do know the flavor of most meats these days is being improved with GMO science. Wonders of nature through modern science." He turned as he walked away.

Jake looked at Charlie with a stern look, "Was that some kind of a slap toward me?"

Charlie laughed, "No, the young man simply said that much of the food we eat is being improved using Genetically Modified Organisms. The cow was fed grain, improved by scientific genetic techniques. The cow is an improved breed, using modern genetic breeding. Science is using genetics to improve our food supplies. Your raw steak may have been on the cutting edge of science along the way."

As Charlie finished his explanation, Jake blurted out. "Oh yea, my inquiry to my friend at Stonemore. He did not want to talk, that was obvious from his text. So, if you want any information, I'll fly over and spend some time with him. If you think it would help her case I can be on the plane tomorrow. I have not had a good English pint in several years."

As they talked of the case, the waiter returned and sat down two cups of coffee. He then laid two peppermint candies down beside the coffee. He also left a black sleeve with the bill inside. "I will be back for this, anything else gentlemen?"

Charlie grinned at the man, "No, I don't think so, my friend is just wanting to sit back and enjoy the afterglow of his delicious steak."

The waiter grinned real big as he said, "He doesn't have to worry about the afterglow, his steak had not been irradiated to kill harmful bacteria."

Charlie laughed out loud and the waiter could not contain a small chuckle; Jake called them both smart asses.

Charlie handed the waiter the bill along with his credit card and then he looked at Jake, "You get the tip, and be generous. You gave as well as you got, besides, you enjoyed bantering with the man. Afterglow… that is a good one."

Charlie and Jake were soon on their way back to their rooms. They talked briefly about what time they would leave for the lawyer's office. It was going to be a long day tomorrow for the both of them.

As Charlie searched his pocket for the key card, he noticed something under his door. It was the corner of a white envelope. He looked up and down the hallway for someone watching as he returned to his room. The hallway was empty.

Charlie slowly opened the door and looked down. Two envelopes were awaiting him. On the outside of one it had his name in typed letters, Mr. Charles A. Morgan. He thought for a moment as to who may know his middle initial in the hotel. On the other was his name simply hand written. He figured it was the information from Nancy he had requested.

He was curious, but suspicious of the letter. He looked about the room and saw a box of tissues. He retrieved one of them and picked up the letter. This may be nothing, Charlie was not going to take any chances.

He felt in his pocket and withdrew a small knife. He laid the letter on the desk and carefully opened it. Inside was a piece of paper folded over in three sections. He opened it, not wanting to touch the paper. As he did, he saw two lines of words. The trouble was, they were not in English. The writing looked to be a Middle Eastern language.

כאשר אדם יוצא לתחום של אלוהים, הוא dams לנצח את נשמתו.
אסור לפתוח את הדלת הבריאה התחום של האדון.

Charlie was puzzled by the cryptic message meant for him, but written in a foreign language. He was becoming more interested in the case, but at the same time that feeling he did not know all that was going on nagged at him.

His mind tried to figure out where to find answers to this message. Whoever wrote it had to know that Charlie did not understand this language. And why write it like this? There had to be a reason, but that reason would have to wait until tomorrow. He knew no language specialist, especially at this time of night. Maybe the doctor would know who to ask. After all, she had traveled the world several times. Along the way she may have picked up some knowledge of Middle East languages.

He then opened Nancy's information and looked it over. He began reviewing the other court rulings and precedence that were part of the brief. After an hour he finished making the changes and made his way down to the lobby. He gave the clerk five dollars to fax the message.

He then returned to his room and took a quick shower. It was late and he wasted no time going to bed.

For a few minutes he stared at the near dark ceiling, wondering what he had gotten himself into. Then he wondered if The General was sleeping under the rock shelf in the canyon beyond the troubles of man? Was his old fish adversary wondering where Charlie was this night too? Sleep soon overtook him and he faded into slumber.

Chapter 12

Come to Jesus

Pre-dawn came early as the phone rang. "Mr. Morgan, you asked for a 4:30 call to awaken you."

Charlie breathed deeply as he said, "Yea, I know, thanks."

He wasted no time getting showered and ready for the day's events. Within a half an hour, Charlie was knocking on Jake's door. He could hear an infomercial playing on TV. It was offering a food processor that could grind a broom handle into dust. In a moment, Jake opened the door and invited Charlie in.

"Sit down, I'll be a few more minutes. I spent time last night on the internet doing a little more research. I used the hotel's Wi-Fi. There is little expectations of security when using a WIFI connection, but at the same time, my inquiries should get lost in all the traffic on a public system.

Charlie said, "Don't care about the internet. What did you find?"

"I went through the back door of a major pharmaceutical company. Took me a while, but finally I discovered the company was interested in the doctors new approach to engineered medicine. One company manager wrote in his internal memo. 'Doctor Mathenson's project has far reaching implication for the future of the drug industry.' He went on to say, 'if it works, today's medicine would become obsolete for those who could afford the new genetic therapy.' His final emphasis was on the incredible bottom line profitability for her discovery. At the bottom of the memo he added, 'We must try and acquire this technology before others find out."

Charlie leaned forward with a questioning look, "They put that kind of stuff in emails?"

"Standard emails." Jake laughed. "No, executive's internal emails. That's where the good stuff is."

Charlie shook his head, "You are going to get us into trouble and Elizabeth already has enough problems without you breaking into the company's private servers. Besides, anything you find out we cannot use."

Jake sat down on the corner of the bed as he tied his shoes. "Lighten up Charlie, they will never know I have been inside the system. Besides, I went into the Commerce Department server and then used it to go inside the drug company's system. When inside, I ran a program interring multiple passwords to gain system access. System security requires that employees change their passwords on a regular basis. Find a low level clerk or associate and there is always one or two that use ABC or 1-2-3 as their passwords. Once I am passed the first firewall, it becomes a little easier as I access basic everyday programs. Daily calendars. Vacation schedules. Things that people don't associate with security. In fact they are meant to be accessible to others. You move from door to door until one lets you in. The trick is then to move quickly, security knew the moment I had entered, and that I was a hacker. But, by the time they reacted, I was in and out. And if they can track the breach, it will lead back to the Commerce Department site. They will deny they had cause to enter their system. At the same time they will know someone hacked in and used their server to gain access to outside files. If it was a breach of national security they would already be here, asking the desk for a list of guests. By process of elimination they would find me. Ain't going to happen, the drug companies won't want to admit I was inside. And the Commerce Department protocol instructs them to contact the Justice Department to show just cause to issue a warrant. They are not going to do that over a simple middle of the night hack with no effort to enter the department internal systems. Those systems are under attack every day from competitors, foreign governments and some kid with an IQ higher than the national debt. Their security people have to separate real threats from simple system probing. They are working to stop all hacks, but that is down the road. With every system fix, someone figures out a new way in. Charlie, this is, or was, what I did for a living. It's like a game. And if you remember, I did it for you and you were part of the government and I was your investigator. I'm hungry, we got time to grab a sandwich on the way into town?"

Charlie just shook his head as he said, "Yes, but we have warrants to look over and the dismissal motion is supposed to be coming from Nancy. We also have to review information pertaining to their investigation, and yes, we'll stop and get you a sandwich on the way. Now hurry up, I want to get into the lawyer's office."

Traffic was still light going into town. As they turned onto the street, Charlie pointed to where the office was located. The early morning street still had plenty of parking available.

Charlie entered the office first. He looked at the clock on the wall, five minutes after six. He heard Downing's voice from the back, "Elizabeth, is that you?"

"No Jim, it is Charlie. Doctor Mathenson is not here yet?" he asked.

Downing came out of the back room carrying a cardboard filing box. "No, she said about six, we'll give her a few more minutes."

Jake stepped in about the time Downing sat the box on the corner of the desk. He looked at Jake and said, "And you are?"

"Jake Meyers, I am an investigator for Charlie and I guess for you too."

Downing looked at Charlie, "You know this case doesn't have the money for a second lawyer, let alone a private investigator."

As the two discussed their involvement, Jake walked to the coffee and picked up the empty pot. He shook it to show to the others it was empty. As he did, he accidently bumped a small container of powdered creamer that fell to the floor. "Shit, no coffee and I made a mess.

Downing looked at the spilled creamer container and at Jake. "Leave it, we don't have time to clean it up now."

Charlie shook his head as he looked at the mess on the floor. He then said to Downing, "I guess you're right… and as to me and Jake's involvement in this case… well, let's just say we are doing this to keep our hands in the game. If I am going to work with you I need him to help with the research. First rule of an embezzlement case is to follow the money. Jake is one of the best in the business. That lets us worry about the legal aspects and the evidence the other side throws against Elizabeth. Jake does the investigation and is paranoid about everything… Oh yes, he cleans up after himself when he makes a mess of a computer file or…spills creamer."

About then the door opened and Elizabeth entered wearing a black suit with her hair arranged neatly on top of her head. She looked professional and ready to face her accusers.

"Did I hear my name?" she said with a bit of a serious smile on her face.

"Yes, I was just wondering where you were. We need to get on the road and into Wheeling by nine. We can grab a cup of coffee and sandwich near the interstate. We can talk on the way." Downing picked up his jacket and laid it over the filing box. He used his foot to pull the door shut as he made his way to Jake's SUV. He popped the back hatch and they loaded the box and their briefcases.

As he closed the hatch, he handed Charlie a folder. "I am guessing this is for you. It's from Nancy. Some old brief concerning a coal company case from years ago. There is also a motion for dismissal for this case."

Charlie took the file and said, "I'll read it in the car. It might have something we can use in this case."

As they headed up the interstate Elizabeth asked, "Just what is going to happen today?" She directed her comment to Charlie, who was just finishing taking a drink of coffee as he was reading his file.

"Today the prosecution will try and convince the judge of the seriousness of the alleged crime. You have been charged with three felonies and one misdemeanor charge. Our job is to keep the proceeding from becoming a trial without a jury. Your job is to look confident and ready to put this behind you."

Charlie turned from his front seat position and looked at Elizabeth, "This is the time you and I have a come to Jesus talk. Do you understand what I am saying? The last thing we want is to get in front of the judge and hear something earth shattering from the prosecutor. Judges want no surprises, especially announcements by either side while in the courtroom for a preliminary hearing, do you understand?"

Elizabeth looked at Charlie with a frozen stare for just a moment before she spoke, "I get your drift from your tone of voice, but I am not sure what else I can tell you about my research or the money. The money is all there, I mean it is all there and accounted for, along with where it was spent. Even the monies from past years. There are no dark accounts. Everything was fine until about six or seven months ago. They started questioning everything. Wanting to know about my trips and if we were close to announcing a breakthrough in my research to be patented by the university."

Downing asked, "So, your telling me they had never paid much attention, at least until about six months ago?"

"I cannot say there was no attention paid to the program monies. But I will admit to spending money that I had not gotten approval from controllers to use ahead of time. Still, it can be accounted for…eventually. Even if it is not proper procedure for a financial expenditure. Quarterly review meetings and reports on the progress of the programs are sent to the grantees' agencies and the drug companies who support our work. Pretty much the same as it has been for the last thirty years. I've never been a favorite of the department controllers. Of course at first it was pretty simple, as was the research. You remember, I told you of those first grant proposals I helped to write twenty, thirty years ago. They were straight forward. Tell them what we want to do and how much money we needed. Truth is, we exaggerated both, hoping to get half of what we requested. But, in the last ten to fifteen years our program along with others around the world, have taken off, biogenetics is the future of mankind. Our research program has some advantages over the others, because we have been going down this road so long. And the truth is, I personally have as much knowledge as almost anyone else in the world. Yes, there are others that have made great breakthroughs, but they each have concentrated on different specific parts of the science. Our program has for ten years tried to bring all those disciplines together in our research. We have made some fantastic discoveries. That's why I have done so much traveling. I shared with other researchers and they shared with me."

Charlie looked at her again, "Don't tell me you are building a genetic elephant in your lab."

Elizabeth frowned as she said, "No, I told you at dinner we are not building or creating a mammoth in the labs. But, we are trying to design a genetic system that can accurately and quickly blueprint a gene. And yes, if someone chooses too, they can build a mammoth or even help a species on the brink of extinction. Rhinos, orangutan, a couple of species of leopards, whales, birds, the list goes on and on. Each year more animals are pushed by man to the brink of extinction. People talk about making a difference, but it is just talk. Maybe in fifteen, twenty years when they get around to making a serious effort to save many species, they will find it is too late. We can at least preserve the genetic blueprint of these creatures. Someday mankind will have the good sense to save them from just being a picture in a museum, to once again be a living animal in the wild. Genetic storage vaults around the world are working to preserve the DNA from many different species."

Jake looked in the rear view mirror at the reflection of Elizabeth's eyes, "Even man? Are you telling me someday you could create a person?"

Elizabeth said nothing for a moment. Then Charlie asked the question once again, "Elizabeth... can your science create a person?"

Elizabeth looked out the window of the vehicle and watched the countryside pass by. Everyone waited for an answer to the question. Finally, after what seemed a long time, she responded. "Do you realize when iron ore is mined from the earth, no one ever asks, what are you going to do with the ore? No one asks, are you going to build things to improve our world. Things like bridges, hospitals, playground swings or homes? They just mine the raw ore and ship it to manufacturers around the world. But, somewhere along the way, one of those manufacturing businesses takes the refined ore and makes a warship or a gun. Those things made from the same iron that is used in the bridges and schools is used to do harmful things to mankind. Does anyone say, 'You should not have mined the ore in the first place and no one would have gotten hurt?' The job of science is to ask, can we understand how something works and, once we do, can we improve upon it. But, once we have those answers there are some that would use the science to do things that could be unthinkable by some. While others want to hold it high as a great achievement of man. Can we create a person?" She looked out the window as they sped along the interstate.

Without looking back she said, "It will happen. Science around the world is prepared to answer that question. The question for the world, what happens when science does?" She turned and looked at Charlie, "Do you think when man first forged steel did he ask himself, will man use this for good or for evil. I have looked in that mirror and answered that question for myself and my program."

Downing quickly asked, "Elizabeth, what is the answer?"

She turned to Downing, "I guess the answer is in the book of Genesis. Once Adam and Eve had eaten from the tree of good and evil, God sent them from the garden into the world. He said to them, man can never return to the innocence of the garden. I can never return to the innocence of not knowing the question of creation for myself."

It must have been five miles before anyone said anything. Finally Charlie looked over his shoulder toward Downing. "We don't know what the prosecution is going to finally charge her with. I had a gut feeling this was about more than money. Especially if she and her staff have found answers to questions of this matter. I am sure there are other players in this game. If the hearing begins heading toward the core of her research and not so much money, we may have to ask for a continuance."

Elizabeth leaned forward toward Charlie and placed her hand on his shoulder. "They want the ADAM Project files."

Downing asked, "Who wants the files?"

"I don't know for sure, like I said this all started about, six, seven months ago. Three years ago I was at a scientific seminar in Edinburgh, Scotland. I heard a speaker from Stonemore speak about a new research program to study the chemical elements of stars, billions of light years away. The program analyzes light as it passes through elements in those distant solar systems. That light then would travel to a new type of photo sensor housed within a special telescope in northern England. The data was sent to Stonemore and analyzed. The new program sensor found almost all the known elements on earth, ninety-four. But, what is remarkable, it also picked up another half dozen elements not known to science. We know about synthetic elements, but these six were new and unknown. Stonemore discovered them, but still has not identified them in the vastness of space. It will take years, if ever, to find the answers to those deep space elements. Some scientists think they are part of the dark matter of space. I don't know about that, but what I did recognize is... if they can analyze trace chemical elements from the vastness of space, can we find a way to pass a light source through the DNA structure of a cell nucleus and capture the chemical spectrum of these structures? We cannot see atoms, but I believed we could see their elemental signatures when converted into light. If this was possible, we then needed to figure out a way to turn that vast amount of information into a map of the chemical elements inside each gene. It took us months to design a method to make it work. DNA is so delicate, almost any light source will damage the gene structure. Light creates heat. Heat tends to cause damage at the molecular level. To find the answers we needed to record the molecule's spectral analysis and then that would help us to determine its chemical elements and how they are arranged within each gene. Finally, we found a way to use light coming from a micro-pulse diode to pass spectrum light through genes and not damage them. The pulse diode, if modulated precisely, would not damage the DNA within the nucleus of the cell. Then we needed a photo sensor that would pulse at the same rate as the diode's light. This step was crucial to the success of the project. We also found a light sensor reflected an almost imperceptible bit of light after each pulse. It was an echo of stray light waves. This tiny bit of light changed some elemental signatures, changing the true chemical analyses. Once we were able to capture the light signatures of the elements, we still needed a computer program to make sense out of the information. I returned to Stonemore and asked Doctor Angus Whitechurch for his department's help. His team already had experience turning raw data into usable information. We began by using the same program used to map elements in stars, only we used light passed through DNA. We hoped it would show us how chemicals arranged themselves in nature to build complex DNA molecules. After a while, the program was able to convert the raw data into information my team could use. Still, it took time even with the power of their massive computer. The results did not come in until about fifteen months ago. That's when things began to happen."

Jake said, "I don't know about the rest of you, but I understood very little of what she said. But given that I don't understand, who else knows about all this?"

Elizabeth thought for a moment before answering, "Not many, our research is compartmentalized to prevent overlap and competition between research groups. Strict protocols are in place to try and keep the programs on track. Even today, I am not sure how many outside the inner circle know the complete spectrum of my research. Each unit is dedicated to its own part of the program. Only a handful of people even know what I have told you today. My research is worth billions to drug and specialized medicine clinics. Our main goal was to find the makeup of DNA, then scan a nucleus that has developed into a cancer cell. If we can see what is normal and then compare its chemical structures to those of the damaged gene, then we can see where things went wrong. Do you know today when a baby is born in most hospitals they ask the parents if they want part of the umbilical cord saved? The T-Cells and genetic coding in the fluid can be preserved and catalogued with this science. Just think, with your original perfect blue print preserved when you are born, it could be used, we'll say at age 35, to repair damaged DNA. Your own body will repair the damage done by time and the environment. This is not science fiction, this is science fact. This knowledge could extend life by twenty years, at least."

Charlie looked at Elizabeth, "I'll ask you this once again, can you create a person from this type of science? Could you have a DNA sequence preserved and not use it to repair someone, but use it to create a person from the very beginning with a single cell?"

Everyone waited for the answer. Time seemed to stand still as they waited. Suddenly the right side tire ran over the rumble strip as Jake's inattention allowed the SUV to drift off the road as he waited for an answer.

Elizabeth finally said, "God gave us the ability to ask about the wonders of the universe. Since they walked out of Eden, our path has led to the point we are at this time. To answer your question, around the world today, not just in my laboratory, scientists are wrestling not with the question, 'Can we do this thing?' they are asking, 'Should we play God and create man?' Yes…my science can. And yes, that is why I think this has all come about. You see, the computer files that have the answers from the last six years were removed from my department's computers. Only three people in my department had access to the entire file. Even though it was on the university's mainframe system, it still needed an access code from me or one of the other two scientists I work with. But somehow the entire file was removed the day I was released from the program. I am not sure if one of the others had anything to do with its removal or if the university isolated the entire system while they figured what to do about me and the research. They still need my access key to get into the files. Even the others do not have the primary access code."

She paused for a moment before beginning. "But, there is a copy hidden away, until answers to some of these questions are found. In fact, I was about to travel to Stonemore later this week to talk with my colleagues about this very question and how they believed the science should be handled. The leaders of the research from all over the world were to gather and try to draft ethical guidelines, to at least make some understanding of where we go from here. I don't think that will happen now. Stonemore is one of the leading universities in both science and religion. If any place may have answers, I believe they will have them. Before long the genie will be out of the bottle in some lab and the world will have to decide what to do after he is out. I know there is no putting him back in the bottle when that happens. You see, I have been indicted for misappropriations of government funds and grant monies, but, I somehow think it is also about my research and my wanting to move the scientific community toward some ethical guidelines. Some don't want to hold back this research. For them it is about money and the future. For me, I want the science to benefit all mankind, just not the ones that can afford it."

She paused for a few moments before explaining further. "You see… I could have made all this go away two months ago when they came and made me the offer to be head of the government's program, with an unlimited budget if I revealed my research. They also carefully indicated that if I refused, my credibility could be destroyed along with my life's work in genetic research. I chose neither, I decided to fight to keep the research secure until guidelines and protocols are put into place so the research will benefit all. And yes, Charlie, this same research can bring back the Woolly Mammoth. Korea is very close, they went another way, but they are close. Do you know that a great deal of their research money comes from genetically cloning rich people's dogs? And there are even rumors that children have been cloned. Just rumors mind you, but money buys many secrets for those who can afford them."

Charlie looked at Jake and said, "Shit, and I was worried about a damned mutant golden trout. Science has now come up with ways to create a cloned person to fish for the damned half breed."

He turned to Elizabeth, "You said the computer files were removed. What do you mean?"

"Some, not all of the files, were removed from the systems. Any files that contained information regarding the ADAM Project were segregated onto a separate secure server. Only a few knew how to access those files. All reports and research were placed in each of the compartmentalized files. If you worked in the…well we'll say biochemistry department. Your access would only be in the bio-chemistry files, but those on my internal staff could access them and move any or all of the information into the ADAM project. If anyone tried to access a portion of the files they were not authorized to see, they were quickly isolated from the entire system. Contact would be made by the computer system to the IT department of a violation. And the file that was attempted to be opened is frozen and only my access could reopen it. It was meant to be like having a puzzle and each piece was in a different room of the building, impossible to do because of the separation of information. Ten years ago, a grad student tried to access the project files. He was just trying to be ambitious and get a leg up on the program to better his position in the research department. That showed us we were vulnerable to problems. I talked with some experts on security that worked. They suggested the same protocols the government adopted at the beginning of the Manhattan Project. Compartmentalization of the research helped to prevent anyone who may be a spy from gaining access to the entire project.

They made the assumption that Russian or German spies were imbedded in the scientific community."

Downing said with a surprised voice, "Spies, I would think you wouldn't have to worry about spies. Sounds to me like everyone in the world is already using the technology. Why steal it?"

Elizabeth paused before answering. Finally she said, "When we started our research years ago, there were those that said we were going down the rabbit hole and no good could come from the research. I don't believe that is true. But, there are people out there who see the technology far differently than mainstream research. Trouble is, they don't have the necessary means to do it on their own. I'll bet in every major research program around the world, there are spies among the programs and working inside the research every day. You wonder why we have been so careful in who we let inside? Just think, what if those who believe in Hitler's master race mentality gained access to this technology in today's world. What would they do with this science? Would they create tall, blue eyed blond haired children? An Aryan nation made in their twisted ideals of perfection? That is possible. No… not possible… it is reality."

For the next few minutes everyone just sat thinking about Elizabeth's words.

As they were nearing Wheeling, Charlie remembered something, "Elizabeth have you had any threats associated with your discoveries?"

Without hesitation she said, "Oh my yes. This type of research has always been controversial. Those that believe man will unleash a terrible plague upon himself when he manipulates bacteria and virus genes. And of course, those who think we are playing God just to prove science has all the answers. But, that perception by some has always been part of the dark side of research."

Charlie reached into his pocket and removed a plastic bag with the note inside of it. He handed it back to Elizabeth. "I got this under my door last night. Not sure what language it is written in. I was hoping since you are widely traveled you might know what it says."

Elizabeth adjusted her glasses as she turned the paper towards the light coming through the side window. "Well, Charlie, it is written in Hebrew. I recognize a few words, but I can't read it. Do you know any Rabbis?"

Downing says with a bit of annoyance in his voice, "Here, let me see that. You two don't get out much into the world do you? He made a couple of swipes across the screen of his phone until he came to an App that was titled, International Translator. He held the note so he could take a clear picture of the words. Then he once again touched the screen and selected the App to translate the words into English. Within a couple seconds it came back with the words. "It's says, 'When man ventures into the realm of God, he forever damns his soul. It is forbidden to open the door of creation that is the domain of the Lord.' "

The translation gave everyone pause for a moment as they absorbed the message. Finally Downing said to Charlie, "It sounds to me like someone sent *YOU* a come to Jesus message."

Chapter 13
Wheeling

For the next few minutes Jake's attention was in finding his way around downtown Wheeling. Morning traffic was heavy and there seemed to be no place to park within easy walking distance.

Finally Jake asked, "Any ideas for parking? I have no idea where people park in this town?"

Charlie pointed toward a private lot, where an attendant sat inside a small booth. "Pull up over there. The worst he can do is tell us no parking."

Jake quickly pulled into the entrance of the lot. He was blocked from entering by an orange swing arm. A skinny man in need of a shave looked disinterested as Jake rolled down his window. The man paid little attention as he was looking at his phone. Then, the man slid opened the ticket booth's side window, and said in a monotone voice, "How long you gonna' need? Dollar fifty an hour. After four it's dollar twenty five."

Jake replied "Not sure, going to the Federal Court house."

The man punched a button inside and a ticket popped out of a slot from a dirty white box. At the same time the swing arm lifted out of the way, allowing Jake to enter the lot. He found a parking place near the corner of the lot.

The group wasted no time heading toward the courthouse. It took only a few minutes to pass through building security and be directed to courtroom one.

A guard stood outside the door and looked them over as they approached.

Downing asked, "Is this Judge Thompson's court?"

The guard shook his head yes and pulled the door open, allowing them to enter. Charlie looked at the clock on the wall. It indicated quarter until nine. Plenty of time to spend a few minutes getting themselves organized before the judge came in.

Charlie looked around the room. Near the front was a court officer stoically looking at those arriving for the day's business. He gave no indication he wanted any interaction. He then looked at the prosecutor's bench. He saw two men and one woman sitting at the table. Behind them were three more men. Charlie was a bit surprised to see his old friend Clay Gordon sitting in the second group. He quickly thought to himself, "*Why would Clay be here and why hadn't he mentioned anything the other day when we talked?*"

As they laid their briefcases on the defense table, Charlie said in a quiet tone of voice, "You see the man in the gray suit…that's Harrison Dobbs. He's been with the U.S. Attorney's office for several years. Spent most of his time working in Washington. I heard before I retired he was looking for a Supreme Court appointment. But, he realized he needed cases with constitutional questions. Especially ones that helped to define his conservative ideals. West Virginia's northern district is close to Washington. Any constitutional case would be recognized by those who make recommendations to the president. I don't think he has much of a chance getting an appointment. In the next few years our president will be appointing moderate judges. Harrison most likely will not be at the top of the list, if he has not changed his ideals. He's a straight up guy, but very conservative in his thinking. I wonder what made him take this case and not assign it to one of the younger attorneys in the district. The other two sitting with him, I don't know, but if they are with him they are hand chosen. Sitting behind them are support people for the prosecution. The older man of the three is Clay Gordon, a senior law professor at the university. I talked with him two days ago and he said nothing about having any part in this. In fact, he gave me the impression he wanted no part at all in Elizabeth's problems. But I have a suspicion he's somehow involved in all of this."

Downing tipped his head toward four men sitting along the back row. One man was obviously a Rabbi. The three others were dressed in dark suits. All four of the men had stern looks on their faces.

Downing asked, "Who are those characters in the back row?"

Charlie thought for a moment before answering. "Not sure, maybe they are some of those people who like to go to legal hearings. But then again, people kept wanting to raise religious questions about this case. Maybe they are here representing God's interest."

Downing said, "Not funny Charlie. But wasn't the note you received written in Hebrew?"

Charlie's face showed a questioning look. "Yes, your right. This can't be just a coincidence."

Jake gently grabbed Charlie's arm. "You remember I told you about the third guy with those FBI agents, well that's him sitting in the back row beside the Rabbi."

Charlie could not worry about who the guy was for the time being. They needed to prepare. He indicated for Downing to place his papers in the lead lawyer position at the table. Then he indicated with a motion of his head toward the prosecution's table. "Let's go make ourselves known and see if they have anything to say before the judge comes in. Watch the younger attorneys eyes… see if anything I say draws their attention. Dobbs and Clay won't tip their hand, but I am hoping one of the others might. Harrison was appointed by the attorney general ten years ago. A simple case like this is outside his normal attention span. That's what's curious about this case."

As Charlie and James walked up to the prosecutors table, Harrison stood and buttoned his coat as he offered his hand to Charlie. "Well, I had heard you were spending your time fishing in the mountains. I am surprised to see you give that all up for time back in a stuffy old courtroom. How did you get yourself into this case?"

Charlie held Harrison's hand longer than the prosecutor was expecting. Finally Charlie released it as he said. "You'll excuse an old fisherman like me, when I get one on the line, I just hate to let them go."

Harrison stared at Charlie with a cold look, "I take it you never heard of catch and release back in your day?"

Charlie said, still locked into a staring match with Harrison. "In the old days before there were so many laws telling people what they could and couldn't do, the decision to release the small ones was left up to the fisherman, not the government."

The young lady sitting at the table stood and said, "I am Penny Winston, I have heard a great deal about you Mr. Morgan. When I was in college we studied several of your cases. I will have to say I learned a lot, especially the importance of paying attention to the details."

For just a moment longer, Harrison and Charlie stared at each other. Finally Charlie turned from Harrison towards the young lady. "I thank you, but I am sure you would have learned those lessons on your own. Before we go any further, I want to introduce the lead attorney in this case, James Downing."

Harrison looked a bit shocked when he heard this, "But… I thought you were here to be Dr. Mathenson's attorney?"

One of the men sitting behind the lead table gave a quick look toward Gordon. Charlie knew that Clay had told the prosecution about his involvement in the case. Charlie began to wonder whose side Clay was on.

Charlie gave no explanation to Harrison as he said, "Well if you'll excuse us, we have a few things to go over before the judge comes in." With that, Charlie and James returned to the table where Elizabeth was waiting.

As they sat down James gave Charlie half a look as he whispered out the side of his mouth, "What was that all about?"

Charlie was already busy writing on a yellow pad as he said quietly, "Never give answers to questions they have no answers to. I know they have talked with Clay and he has told them I was interested in the case. They just can't figure out why I would come out of retirement to defend someone I don't know. A lawyer never wants unanswered questions getting between them and their case. Every distraction you can create for the other side will help your case. If I can create a question in his mind, he will wonder if I can do the same thing in the courtroom when the jury is here. Harrison will not be deterred by my answer, he has already moved on. But, he will have his investigators try and find the link that ties me to this case. The other attorneys will have some doubt about me. And you heard the young lady, she already thinks she knows how I think from reading old case files. That's a good thing, now let's get down to business."

Over the next few minutes, Charlie and James went over the case. They discussed how to handle the questions of evidence the government was going to offer to the court. And they wanted to keep Elizabeth out of jail until the trial date.

The three were busy looking at the evidence and talking about what the prosecution may present to the judge.

As it neared ten o'clock, Charlie saw the bailiff open the door into the judge's chambers. He knew he was checking to see if the judge was ready to enter the court room. In just a moment he stepped back in as he said, "All rise, the Honorable Senior Judge Harold Thompson presiding over the Northern District Federal Court."

Thompson was a large man in physical stature. He was not a heavy man but he looked like he could have been a formidable lineman for the Steelers in his younger days. In his court robe, he appeared even larger and menacing to those who sat before him in his courtroom. His snow white hair struck a sharp contrast to his dark robe. There was no doubt he loomed large in his courtroom. In a strong deep voice he began, "Alright, let's get on with this. You all may sit down."

He picked up his glasses and placed them so he could read documents on his desk and at the same time read the faces of those before him.

"Mr. Dobbs, I see we have an indictment of Doctor Elizabeth Mathenson. You have charged her under Section 18 U.S. Code 665 – two counts of theft or embezzlement from employment and training funds; improper inducement; obstruction of investigations. A third count of wire-fraud in regards to solicitation of funds not within the jurisdiction of her research. And the fourth, misappropriation of funds not in accordance with purchasing of items not authorized by the program guidelines."

He looked up at the prosecution table, "Mr. Harrison, is this the section you have charged Doctor Mathenson with violating at this point?"

Harrison stood and said, "Yes your honor."

The judge wrote something on his papers before he looked up at the defense table. "Doctor Mathenson, you have counsel for your defense I take it."

Elizabeth stood as she answered, "Yes, your honor, I do."

"Do you fully understand the charges that have been brought against you by the government?"

Before she could answer, Downing asked, "Your honor I understand the government has charged my client under section 18. That implies embezzlement or theft from employer. It is our understanding the case against Doctor Mathenson is embezzlement of research funding. Does the prosecution intend to bring an all-encompassing indictment? Blanketing my client and making it impossible to answer specific charges regarding the alleged missing grant money."

The judge sat for a minute looking at the documents before him. As he read the indictment, Harrison stood to say something. Thompson held his hand up, indicating for him to hold his comments until he read the indictment once again. After he completed his review he looked at the prosecutor, "Yes, Mr. Harrison, you were going to say something?"

"Your honor, it is the government's contention that Doctor Mathenson has for many year's diverted monies from well-defined federal grant programs. She has redirected those funds to programs not approved by her colleagues or the university. Doctor Mathenson has, without regard for proper documentation or proper oversight, spent and defined funding for projects not approved by the university's internal review panel. This failure on her part has led to loss of funding. The governments intends to show Doctor Mathenson committed a felony under section 18 of the code. Doctor Mathenson has fraudulently redirected funds earmarked for other research departments. We believe this constitutes theft from her employer. It is also important to note, that when her actions were questioned, she removed information from the university's computer files to hide her illegal actions."

About that time, Clay leaned forward and whispered to Dobbs. "Don't forget the question of research specimens and their whereabouts."

Charlie moved his hand toward James who was already rising. He understood Charlie wanted him to challenge Harrison's last statements. "Your honor, the prosecution is trying this case already and the jury has not even been empaneled. His unsupported charges would better be served before a jury who would quickly realize he is spreading the government's charges very thin, hoping something sticks with his alleged theories. The question before this court has to be, 'Did Doctor Mathenson improperly spend funding allocated for research?' We intend to show she has directed her actions for many years with the same regard to protocol when it comes to funding. Are we to believe, the university has suddenly deemed her work not to be part of the original project? We therefore are prepared to submit a motion for immediate dismissal of this case on grounds she violated no laws regarding the direction of her research. This is a matter internal to the university and not a violation of federal law."

Harrison stood and objected to the statement and the motion for dismissal. The two men for a moment were engaged in a verbal contest of words. Finally, the judge struck his desk sharply with his gavel.

"Gentleman, we are here today to set a trial date and ask the defendant a question. As to your motion for dismissal, I will review it, but I can see no reason that this case should not go forward at this time. If I change my mind I will advise counsels, but that is unlikely"

The judge looked at Elizabeth, "Doctor Mathenson, I ask you once again, do you understand the charges and how do you plea in this matter?"

Elizabeth looked down at Charlie who nodded for her to proceed, "Yes, your honor, I understand I am charged with embezzlement of funds used by my department and I plead, not guilty to all charges."

The judge sat for a moment, then asked the two attorneys to approach the bench. "Gentlemen, my docket is pretty well open since I have been pulled out of retirement to hear this case."

He looked at Harrison, "When will the government be ready to proceed?"

Harrison, with a bit of hesitation, said, "Your honor… we can proceed at your discretion."

The judge then looked at Downing, "Son, just how many federal cases have you handled?"

"None sir. And if I may ask, how does my time in a courtroom have a bearing on how you will proceed in this case?"

"Careful, Mr. Downing… my questioning your time in court is well within my scope of concern for your client and the integrity of the laws, we are here to administer fairly. You are not known to this court. And your client is entitled to a proper legal defense. I'm not questioning your dedication, I am however wondering about your experience in a courtroom when it comes to a complex felony financial embezzlement case. If you fail to understand the charges and the statutes sited by the opposing counsel, you could miss proper rebuttal of the point. Remember, the members of the jury have less experience than you. They'll not know if either party misses a point of evidence that may sway them one way or another. I'm just hoping your lack of experience is not an attempt to at some future date, to set up an appeal for inadequate defense on your client's part. A new trial granted on the grounds of an inadequate defense also reflects on the judge in a case. I have never in all my years on the bench had a new trial granted on those grounds. Do you understand, Mr. Downing?"

"Yes your honor."

Then the judge looked toward Charlie. "Mr. Downing, where did you dig up an old U.S. Attorney like Charlie Morgan? I take it he is going to be assisting you in this case?"

"Sir, Mr. Morgan has been engaged by Doctor Mathenson and yes, we will be working together in defense of her case."

The judge sat for a moment thinking before asking Downing, "The prosecution is prepared to begin their case, when will you be prepared with your client's defense?"

"Your honor, before I can answer that question, I need to know when the prosecution is going to present its evidence for our review. So far his office has only provided an overview of evidence. I may be inexperienced in the court room, but I know we have the right for pre-disclosure of evidence that the government is using to support its charges against my client. Not only has the prosecution applied the statues widely, he has provided little in supporting evidence of the government's claim. Your honor may be better in questioning the government's intent on prosecution in this case. I see little evidence that even a relatively new attorney like myself would need to question the extent of the case and supporting evidence."

The judge turned to Harrison, "The young council seems to understand evidence and the government's responsibility for pre-disclosure. Well, Mr. Dobbs, has the prosecution not provided the defense with the evidence they intend to present? I cautioned Mr. Downing on not basing his case on the assumption of a mistrial or re-trial. I wouldn't think I would have to make the same point to you."

"I am not sure what additional evidence Mr. Downing is referring to, but in an effort to move this case along, I will ensure my office will hand carry the pre-disclosed evidence this afternoon to his office for pre-trial review. I would remind the defense, it is the government's responsibility to provide him with the evidence. He is responsible for any future investigation as to his client's defense regarding the evidence."

The judge looked back at James, "Anything else Mr. Downing?"

Downing shook his head as he said, "No, your honor, at this time I see no reason why we cannot begin with our defense in one month. At that time, we will be ready to proceed with my client's defense."

Harrison nodded his head in agreement and said, "That is acceptable to the prosecution your honor, but there is one more question."

The judge looked toward Charlie for some indication he approved of the time table Downing had put forth. He gave no indication he was unhappy with the trial date being set in one month. The judge then motioned for the attorneys to return to their tables.

"I assume the prosecution has a recommendation for the defendant in this case, as to the question of bail?"

"Yes, your honor, we believe Doctor Mathenson is a possible flight risk. In fact over the last few years she has traveled out of the country on many occasions. South Korea, England, Israel, Africa and Scotland are a long ways from a West Virginia courtroom your honor. Doctor Mathenson, I would assume because of her travel portfolio, is at ease in leaving the country at will."

The judge looked over at the defense table as Charlie stood, "Your honor, Doctor Mathenson is eager to put this behind her and move on with her important work and restore her position at the university. She has no past history of illegal activities that would indicate she is any type of flight risk. Her home and family, along with her career, are in Monongahela City. She will not leave the country and she is prepared to surrender her passport as a showing of her wanting to get on with this case and put it behind her."

The judge stared at Charlie for a few moments before saying, "Mr. Morgan, as warm and caring as ever I see. Did you catch all the fish and return to the courtroom for something to do?"

Charlie chuckled for a moment before saying, "Well your honor, I guess you could say I miss the excitement of the courtroom once I heard you had returned."

The judge grinned at his answer. "Doctor Mathenson, you will surrender your passport and bail is fifty thousand dollars. You are instructed not to leave the state for any reason without contacting the court. Is that very clear?"

"Yes, your honor."

For the next few minutes the judge was busy writing on the papers in front of him. Finally, he looked up and said, "I will hear the case in thirty days. That puts us back here on Monday, the twenty-third at one p.m. Is there anything else either party has to present to the court at this time?"

Both councils responded, "No, your honor." With that the judge struck his gavel and dismissed his court.

Charlie stood and looked at Elizabeth, "You'll need to go with the bailiff and surrender your passport and make arrangements for your bail. I assume you can come up with the ten percent to secure your bond?"

She indicated that she was prepared to make the arrangements as the bailiff stepped up to the defense table. "Doctor Mathenson, you'll need to come with me."

As he finished he looked at Charlie, "The judge would like to see you in his chambers."

Charlie glanced over at Downing as he said, "Don't know what he wants, but I'll go see." Downing's expression looked as if he was the kid left unpicked for playground games. Charlie closed his briefcase and headed toward the judges chamber. Elizabeth followed the bailiff to arrange her bail and surrender her passport.

From behind him, Downing heard Jake's voice. "Not bad, not bad at all. You say you have no experience in the courtroom?"

For a moment, Downing paused to reflect on Charlie's leaving before answering Jake's question. "No, not really, I have attended dozens of real courtroom cases watching how lawyers prepare themselves before judges and juries. And I have taken part in several mock trials in college. But, this is the first time I have stood up and defended a real case."

Jake chuckled, "Well, for one thing Judge Thompson is a decent guy. He cut you some slack on the evidence question. But, you also pointed out that the prosecution was trying to argue the case before the trial has even began. You did okay. In fact, you got in some of your own argument."

Downing looked toward the door where Charlie had gone. "What's that all about? Can the judge just ask for counsel to come to his chambers to discuss the case?"

Jake looked toward the courtroom doors and the bailiff who was waiting for them to leave. "There is a coffee shop downstairs, let's go get a cup and wait for Charlie and Elizabeth down there."

The two men made their way out of the courtroom and down into a small coffee shop. After they sat down Jake said, "Charlie and the judge have known each other for years. They'll not talk of the case, at least not in direct terms. I would suspect, it is two old friends catching up on the last few years.

Charlie knocked on the door of the judge's chamber. He heard from within a voice that said, "Come on in."

Judge Thompson was removing his robe as he said, "Well, Charlie, how have you been? Since we both retired I never get to see you anymore. Did you ever catch that fish you were always after, what was it you called him, The General?"

Charlie chuckled. "No, not yet, but I am still trying. With my knees, I have a tough time making my way into the canyon these days. I don't think there will be too many more times I can go after him. I may just have to settle for the one that got away. How about you, what have you been into since retiring?"

"Baby sitting mostly. My son and his wife both went back to school and we have been helping out with the kids. Keeping up with preschoolers has been a full time job. They don't care if I am a senior judge. To them I'm just Papa."

Charlie sat for a moment, "What did it take to pull you away from those grandkids and back on to the bench?"

"The sitting district judges all have cases pending. Two have been delayed for further review by the justice department. And Peterson took a medical leave. Didn't hear what was wrong with him. So, I guess the justice department decided, rather than have the judges balance several cases, they asked me to come in and preside over a couple cases. I'll have to say, I never expected to see you in the courtroom, especially on the defense side of the room. I always believed you were born a prosecutor and had no interest in the defense."

Charlie thought for a minute, "Well, I guess in some ways I am as surprised as you are. But, I'm finding I kinda' like getting away from the house and office. I spend one day a week there, mostly doing wills and deed transfers, things such as that. Nancy has been with me for years and she is one of the best researchers around. I always hated sitting in a basement somewhere reviewing old deeds and record books. Nancy has run my office and been my eyes and ears in my practice. I guess in some ways I figured to retire and work a few days a week, while she keeps the office operating. That way I could keep her working and my retirement plans alive. Lately, she's been talking about retiring to Florida. Her son and his family live near Jacksonville. Maybe, if she goes I will finally give it up completely and stay home and practice my fly tying skills."

Both men sat for a moment not saying anything. Then Thompson said, "Is the kid up to the job of defending the doctor?"

Charlie thought for a moment, "Yea… I think he is. He's quick and knows the law. More than that, he knows the parts of the law that are between the lines, the intent. He'll be okay." With that Charlie looked at his watch, "Well, I better get downstairs and see if Elizabeth's made bail and the others are waiting." He stood and shook the judge's hand.

As Charlie stepped into the hallway outside the courtroom, he saw Harrison standing, waiting for him.

"I should be talking with the lead attorney, what's his name, Downing? But, I want to run something by you before we go further down this road. The governments is willing to reduce the charges to simple funds misappropriation with no intent, if Mathenson will return information and disclose the location of a missing cryogenic chamber we believe is in her possession. She would also have to explain questions in the funding of a project known as ADAM. Next, she would have to allow complete government oversight of her project. The oversight would be administered by Clay Gordon. He would assure that both the university and the government's interest are protected. Finally, any travel the doctor does will require prior approval and she will be accompanied by a member of the oversight committee. Charlie, I can't say I understand this biogenic science stuff, but, I do know it has potential for a great many things. Mathenson has managed to keep the information compartmentalized. But those working on the project realize she is on to something groundbreaking. Then add in the travel to countries like South Korea and Africa. Now, I have also learned she spent time in Israel. For some reason I have not figured out, their government would like to meet with her. In fact, they are insisting on speaking with her. All of these countries have advanced research centers. Is she working for the university or … well you can see my point. My case is only about the money, but if she is willing to work with the people I mentioned and follow these new guidelines, I think we can go back to what we were doing last week. You can catch that fish you always talked about. Oh, one more thing, this would become a black government project. No discussion by her or even your team of the nature of the offer or what you have learned about her research project. The government believes this has great importance for many different reasons. Topping the list, national security. And for reasons I've yet to understand, the Israeli government is also pushing their national security issue in this case with our state department."

Charlie was a bit stunned by Harrison's comments and offer as he said, "My client is on trial for embezzlement…plain and simple. But, you make me think there is a whole lot more than just those charges. You realize her tenure, her reputation and worldwide credibility has been severely damaged by the government's charges. I damn well think you could have come with an offer a month ago before the grand jury indictment and saved a lot of trouble for everyone. Why now? And why didn't you bring this up a few minutes ago before we went before the judge?"

Harrison's face grew cold, "Charlie, I have, well, let say ambitions. That has never been a secret. I was asked by the attorney general personally to handle this case. Get a conviction if necessary, but someone in the government is more interested in the research your client is doing. Rumors are that your client went down a rabbit hole with this research in countries and places that may not have the same security interest as we do. You tell her if she wants to follow her research as far as possible, all she has to do is play ball with the university and the government. She will still have a free hand in her research, but with oversight. But, this offer goes away if we go to trial… is that clear?"

With that, Harrison smiled and turned and walked down the hallway. After a dozen steps, he stopped and turned back toward Charlie. "My name is said to be on the moderate short list for the next Supreme Court position. I have no interest in going to trial and have your client go on TV and tell how the government is prosecuting her for her beliefs in science. Get her to take the deal."

Charlie had not noticed, that during Harrison's final words, his two assistant attorneys came into the hallway. And even more curious, behind them, Clay Gordon was talking with the four mystery men who were seated in the back of the courtroom, during the hearing. He felt he was being played by Harrison. But, someone else was pulling his strings. Charlie felt his ass pucker. He somehow felt he was getting the shaft and he didn't even know why. And what the hell was Clay up too? But, one thing Charlie was sure of… he was going to find out.

In a few moments, he was going down the steps looking toward the coffee shop. He could see Jake and James sitting at a small table. He didn't feel it was time to let them in on Harrison's offer here in the open courthouse lobby. He would wait until they were heading home.

As he approached, James looked up at Charlie. "Well, did you and the judge have a good conversation?"

Charlie laid his jacket over the back of the chair as he sat down. "Just two old friends catching up on the last few years. And no, he did not say anything about our case. He knows his role in the courts and he would not ever taint a case. He did how ever ask my opinion of you as an attorney."

Downing seemed to bristle at the notion of being reviewed by the judge. "Well, what did he say?' he asked sharply.

"He simply asked if you were up to the case and I told him I had every confidence in your legal abilities. We also talked about him becoming a full time babysitter of his grandkids. Don't worry, just two old friends saying hello."

About then, Elizabeth came from the second floor holding a handful of legal papers. As she approached she said, "I have been through a lot of things in my life, but that is the first time I have experienced anything like that. I am ready to get out of this place."

The others quickly agreed and soon they were on their way back to Monongahela City. After a while Downing asked Charlie, "I was correct in saying the prosecution was casting a broad net in the indictment…correct?"

Charlie, who had said very little while looking out the window of the SUV, turned toward Downing. "Yes, you were correct, and that is what's bothering me. Why would the top prosecutor in the eastern district be asked by the US Attorney General to handle this case? The indictment is broad in scope and Harrison made no attempt to define the government's case when you raised that question. And another thing, he was not as aggressive as I have seen him before. He realized you were inexperienced and could have used that to try and intimidate you in front of the judge. He didn't. Why? He went out of his way to stay on the straight and narrow of the embezzlement charges. Harrison is one of the top litigators in the justice system. He comes straight at you because he's good. He prides himself in the belief in the law and the law will ensure justice. And why did he have Clay Gordon there? Clay never even made eye contact with me when I was only eight feet away from him. I've known him for years and he is also a longtime friend to Elizabeth. His being there is not right. I'm guessing the university has more than a passing interest in this case. Clay is an expert in patent law and written extensively about the process. One more thing, the Israeli connection. It seems to keep popping up."

Charlie paused for a moment and then instructed Jake to pull off at the upcoming roadside rest area. When the SUV came to a stop, Jake turned off the engine. "All right Charlie, what's this all about?"

Charlie repositioned himself to be able to look in the backseat more easily. "Harrison approached me after coming out of the judge's chamber. He wants to offer Elizabeth a deal."

Elizabeth quickly perked up and said, "Deal, what kind of deal?"

"He will reduce charges to a misdemeanor charge of misappropriation of funds. I'm guessing that is negotiable too if you're interested. But, that offer comes with a set of conditions. First, you must return all information and files associated and pertaining to the ADAM Project. Also, he says they believe you took some samples and some kind of cryogenic chamber. They want you to explain all inquiries and questions concerning the project. Third, allow for complete government oversight of the project. The oversight would be under the supervision of Clay Gordon. Finally, any travel out of the country would require prior approval and someone from the oversight committee would accompany you. Do this and the charges would be reduced. You would also still be project manager, but with oversight from Clay. One more thing, and this affects all of us, this whole affair is to be classified under national security guidelines. In a sense, if you agree we all are bound by the terms of the settlement. He also says the Israeli government wants to speak with you."

Elizabeth, who was listening to the conversation said without hesitation. "No! No damn way am I going to turn my life's work over to the government. And besides, I want this research to benefit science and medicine, not to be used for secret government projects. I told you this has great potential in many areas."

She looked a Charlie, "Aren't you the one who said this is just about the money. I tried to explain it was more than the money. You told me it is only about the embezzlement charges, but what if it is being driven by my research and the possibilities it raises. I believe if you are going to defend me, you have to open yourself to the possibilities it is more about the work. For some, bio-genetic science violates not only Mother Nature's laws, but mankind's beliefs where we came from. I have no doubt that

Jake's audit of the grant programs money will reveal it is accountable. Maybe not totally approved of for some of the things I used it for, but it is there. But, his audit will not answer the real question… where has my department's research directed the money to be spent. Money was not the big issue by auditors in the last year, it was more the direction and scope of my research. Now, I know… it might be violating some higher laws, but the government wants the science… no matter what laws it might be violating. Questioning the use of the money by me is only a way to get to my work."

Jake looked at her as he said, "Higher laws, you mean higher than the federal courts?"

"Elizabeth paused before she answered. She leaned forward toward Charlie, "You remember your note written in Hebrew. What if this is about the law of creation? Do you remember that first evening we met? I asked if during the discovery of evidence in a case, have you ever learned a truth that not only changed the outcome of the case, but altered your life? A truth bigger than the case itself?"

Charlie looked at her with an odd expression. "No, not really. Federal cases are pretty much about the evidence and I've seen it all. No bigger issue than the evidence. Besides, I am not sure what you are talking about. Are you telling me you are hiding something? Remember I told you up front I need you to be honest with your defense team. Secrets are a sure way to end up in jail. Now, what are you referring to?"

Elizabeth sat looking out the window not answering Charlie's question. Finally she said without looking at him, "You said it is all about the monies the government alleges I embezzled. I have told you everything I know about the money. And I have told you, my research is about genetic medicine… to find answers at the most basic levels of human biology. As to a deal to give them my life's work, no, I do not want a deal. I want to complete my research with no oversight. Just let them try and convict me in an open court. They won't do it. I now know they are not after me, they are after ADAM. And they won't try in front of the world to reveal its secrets or mine."

Elizabeth still never looked at Charlie. In the silence of the car, Jake could tell Charlie was pissed at her words. He was convinced Elizabeth had some bigger secret she was not telling them. For a moment Charlie thought about saying "the hell with it" and returning to the mountains. His client was keeping secrets. The prosecutors lack of aggressiveness in the case. Clay showing up in the courtroom. And Israeli agents showing interest in a case far outside their normal channels. Charlie began to feel like he and Downing and Jake knew less about this case than anyone else associated with it.

Finally James asked, "I know this is all about the money, but what if it is really about this ADAM Project. And by the way, just what is the ADAM Project in simple terms?"

Elizabeth sat for a moment collecting her thoughts. "ADAM is the name of a project that is being used to create altered DNA. To repair damaged DNA in a child born with Down syndrome or a genetic blood disorder. We can identify the damaged gene before the child's birth, remove a sample and repair it, reinsert it into the body and hope it begins to repair other damaged genes."

James perked up, "Forgive my ignorance, but that's not new science. You were experimenting with that even when I was in your lab five years ago. And I see on TV where they are doing a variety of gene therapies around the country. What's so special about the ADAM Project's work?"

Elizabeth looked at each of the people in the SUV. "The ADAM Project has been designed to replicate dormant DNA strains."

The technology of ADAM can synthetically stimulate the genes to rebuild themselves. The possibilities are endless. And if the science were to get into the hands of a drug company or bio-engineering firm, it would be worth billions. No, that's not right, it would be priceless. Even more than that, it would change man as we know him. I asked you if you believed in God. And if you do, then you believe God created Adam. But what if the science of my ADAM Project has the potential to create……"

Jake interrupted, "Create what?"

Elizabeth answered, "Create what is believed to be only possible by God's hand. A human person."

Chapter 14
The Reveal

No one said much as they traveled along the interstate highway heading south. Charlie understood what Elizabeth was trying to explain. But, he was also confused by the complexity of her answers and his lack of scientific understanding. He thought to himself, *"this is a little too fantastic to be true in some ways. If this science is out there, why haven't others made the same discovery?"*

He turned to Elizabeth, "Why your project. Surely others have found the answers to DNA's dormant code. I heard on the Science Channel about dormant DNA. Science must know the potential of the dormant strains?"

"That's true and in another six months, a year, someone else will make the same breakthrough I have. I discovered something in my travels around the world that they have not yet realized. I watched as dozens of pieces of the puzzle were being researched and tested by brilliant doctors and scientist. Each of them was too close to their work to see the bigger picture. After a while I stood back and could visualize something the others had not seen after a trip to Korea. That's when I remembered something in a scientific paper written by an engineer I knew a long time ago at Stonemore. That discovery changed everything."

James asked quickly, "What was that?"

Elizabeth said, "Well, I guess that is the sixty-four dollar question isn't it? And if the case against me is about money, then my discovery should not be an issue in my guilt or innocence… and I am innocent of the embezzlement charges. Anything outside that is none of anyone's business. And that is why I am not going to accept the prosecutor's offer. Let him make his case in a public courtroom. If he does, the world will know the secret."

In the silence Charlie and the others were trying to figure out the turn of events that day in light of Elizabeth's revelation.

Finally nearing their exit into town, Elizabeth pointed out a large advertising billboard in the far distance. She tapped Charlie on the shoulder as she pointed toward the sign.

"Remember our unexpected visitor at the lodge?" Charlie, who had been in deep thought and was in no mood for conversation, had not taken notice of the distant billboard until she pointed it out to him. It was a large advertisement for the Church in the Sky. A twenty foot high image of Pastor Jedidiah Cain's head and shoulders. It was inviting those lost in today's world to attend his services and find their way home to God.

Charlie looked at the sign as he said grudgingly, "Hell, maybe I should go ask him for answers about this case. Can't hurt, I'm not getting them in this car. He certainly seems to have them." He grumbled some angry words that the others could not understand. But, it was clear he was not happy about the circumstance he found himself in. In fact, he was downright pissed off at the doctor. He knew she was withholding information. And she seemed to avoid all together the question of the missing DNA samples and cryogenic chamber. In his mind he knew if she had taken these things she was not the victim she let on to be. She had a hand in the whole affair. She was guilty, but of what?

He looked up at the rear view mirror and could see Elizabeth's eyes in the reflection. In a firm tone of voice, he said, "I don't like feeling I am not being told everything when working on a case. AND… it is important for me to decide what is and what isn't important!"

With that comment he continued to look at Elizabeth until Jake said, "I'll need to make a few calls and spend time on my computer doing research. The final records the prosecution gave us need to be analyzed and gone over. It will take me a few days to get a handle on the numbers. I'll go back ten years. That should be enough to use in her defense. Don't think anything before that would be pertinent."

Jake's comment was as much about breaking the tension of the moment as telling the others what he intended to do next.

Downing seemed to have the same intent as he said, "I have two spare offices that are being used for storage until I get set up. You are welcome to use one and do your work."

He tapped Charlie on the shoulder, "And I guess since we will be working together you can use the other office. It's not much, but it will allow you to work and use a computer for research."

After that, the silence in the SUV was so thick you could cut it with a knife. Elizabeth's revelation about the case made everyone uneasy as to what they were getting into.

After twenty minutes the group had made their way to a parking place in front of Downing's office. They each stretched their legs as they exited the SUV. Jake popped the tailgate and Downing and Charlie each picked up legal boxes and headed for the office door still not talking with each other. As they stepped up on the sandstone steps Jake stopped abruptly as he looked down at the steps.

"Look there, what's that white powder on the steps coming out of your office?"

Charlie looked down at the steps and said, "What the hell are you talking about? I don't see anything on the steps. Let's go, I am tired and I want to get back to my room."

Jake adjusted the box to free one of his hands as he pointed to an almost imperceptible smudge of powder."

"There... that was not there when we left."

Charlie pushed by Jake as he said, "Bull shit, I've told you that paranoid attitude will get you into trouble one of these days." Charlie grumbled as he pushed open the door, "Burner phones, men in black suits, monitoring phone calls, my client's secrets...and now... powder smudges on steps. This is a red letter day for shit going wrong. Jake you should have been a mystery writer. But don't ask Elizabeth for a story line, it's a secret. You see any boogeymen hiding in the office Jake?"

Once inside the office, Jake quickly went to where he spilled the powder creamer earlier in the day. It was gone. Jake sat the box down and quickly knelt down to the floor to a position in which he could look across the floor in an almost parallel view.

He turned to the others and placed his finger in front of his lips. Charlie started once again about Jake's attitude. Jake placed his hand in front of Charlies face. Jake shook his head no as he indicated for the group to exit.

Once outside and part ways down the step, Elizabeth said, "What is going on? Why have you brought us back outside? It's been a long day, I want to get home."

Jake looked around the street before he answered. "Alright, look here, my job is to investigate and be paranoid. And see things on the dark side of the law. And I'm telling you something is wrong in that office. Someone's been in there since we left."

He pointed to the almost imperceptible evidence of coffee creamer on the steps. "Someone stepped in the powder in there and tracked it across the floor toward the filing cabinets. It must have been too dark for them to see the spilled powder when they first entered. But, somewhere along the line they discovered their mistake. They could leave no evidence of their being here. "

He stood up and looked at the group, "Someone was in this office after we left. They had no choice but to clean the floor. They knew what they were doing and cleaned it spotless, almost. They figure if no one saw it when we returned, it's unlikely anyone would even remember the spill. But, that old rough flooring trapped a few grains of creamer."

Elizabeth said, "Are you sure, maybe James had someone come in and clean." She turned and looked at James.

"No, there should not have been anyone in here this morning. I can barely afford the rent, let alone a cleaning person. I didn't see any tracks to the filing cabinets, but I am sure he spilled the creamer before we left. He's right someone was in here, not sure why."

James turned to Elizabeth, "See, this case is becoming more complicated by the minute. You are the only one that knows what is going on. This has to do with your case. There is nothing in this office that would be of benefit to anyone in getting information about this case. We had everything in the vehicle."

Charlie stepped toward Jake. "Alright, someone was in the filing cabinet… which of the filing cabinets did your mystery person look in?"

Jake opened his briefcase and fumbled around looking for something. Finally he said, "Ah-huh, found it." It was a small flashlight, the kind Special Forces soldiers carry. He turned to the group, "Wait here for a minute while I go see if I can find which cabinet they looked into."

Jake stepped inside the door and switched the overhead lights off. Next he turned the small flashlight on and shined it into the darkened office. Carefully he pointed it toward the center of the room as he knelt down onto the floor. Focusing the lights beam to a narrow sharpness, he laid it on the floor shining it toward the cabinets. Jake adjusted its position until he saw what he was looking for. Standing, he placed his finger against his lips, indicating to the others to remain silent. He then pointed toward two areas on the floor. His gesture meant he found evidence of the creamer where it should not have been.

Charlie turned to Downing and gently took his arm leading him away from the open door. Then in a hushed voice asked, "What's in that second cabinet?"

"Nothing important, just old cases I researched in college. You probably remember how the professors would give you the case evidence and then have you write a brief summarizing it. We were expected to draw our own conclusions as to where the evidence points. After writing the paper we were given the attorney's briefs from the actual case files. That way we could read what the attorneys in the case wrote in their summation briefs to the judge. Charlie, there are a couple of your cases in that file I wrote briefs on."

Jake picked up his flashlight and made his way outside to where the others were. He asked Downing, "Do you think you could find which file was looked at or taken?"

Downing thought for a moment and snapped his fingers as to something he remembered. "Yes, maybe, I catalogued all my briefs in order to be able to look them over if I ever got a similar case when I went into practice. If they took the file, it will be easy. If they only looked at it…I am not sure I can tell that."

Elizabeth had stood quiet for the few moments, "Alright someone was in the office, but why are we still outside and not going in and calling the police?"

Jake was quick to answer. "If when I was investigating and went through all the problems of picking a lock and entering an office, I would have photographed what I wanted and…and I would also have left behind a bug." Elizabeth looked oddly at Jake.

"I would have left a listening device. Any good investigator would do the same. Small inexpensive and untraceable. I'll bet somewhere in there is a device to listen to our conversations."

Jake pulled Charlie aside, "This is getting weird and serious. The government wouldn't do this. Not for a simple embezzlement case with a green kid lawyer. Someone else is wanting to be part of this and keep their identity quiet. You better find out from the doctor what she has gotten us into!"

Charlie thought for a moment, "Let's get a cup of coffee a couple of streets over." Elizabeth and James were a bit confused as to the sudden need for coffee and Charlie's seeming to disregard that someone entered Downing's office, but they agreed.

As they got into Jake's SUV he turned to Charlie, "Watch and see if any vehicles on the street follow us."

Charlie watched to see if any vehicles followed them around the corner, but he saw no cars or trucks that stirred his suspicions.

In a couple of minutes they entered the coffee shop and found a small table in the back near the wall.

Charlie began, "Jake, I want you to start your research into the dark area of the information web? There is nothing about this case that seems normal for a case of embezzlement. Try and find out if there is any chatter about possible breakthroughs in biogenetics. And look into Clay Gordon's background. He's a player. I don't know what his involvement is, but he's up to something. See if he's something special with the university science board or even a consultant for the government. And see if he is going rogue on his own. If he senses a major patent discovery, it could be his ticket to the good life. If someone at the university is after her discovery and Clay gets a fee for bringing the patent home to them, we are talking millions."

He then turned to Elizabeth, "You have to know something about what's going on with Clay. You suspected someone for nearly six months.....was it Clay? I know you asked him some questions last summer about patents. You know something...WHAT!" He shouted at her. "If Jake's right about the office and possible listening devices, you're in deeper than the case the governments got against you. I'm a simple old prosecutor, but I know when things get weird. I once worked on an international case with embezzlers who worked with smugglers. The CIA was involved in that and they were a sneaky bunch. This smells of that kind of case. You talked about a billion dollar discovery. People will kill for a few thousand. They will do anything for this kind of money."

He shook his finger at her, "I WANT TO KNOW WHAT'S GOING ON!"

Elizabeth was startled by Charlie's outburst, "I've told you everything I know. Go talk with the science board. They changed their attitudes toward my research in the last six months. They brought lawyers in to review the books a couple of times. Questions of the project's goals by the board over the years is not unheard of…and even having an attorney in the meeting is not totally out of line. But somehow the meetings were proper and formal. I mean there was not small talk before or after. They asked questions and I gave answers. Everyone was polite and stayed on issues before us. Go ask the head of the board; she can give you some answers."

Charlie looked out at the busy street. He stared at the cars and trucks. After a couple of minutes he turned back to the group, "I am getting as paranoid as Jake and I don't like that. We'll work the case. Jake do your thing with research. But, see if you can find any information about the head of the science board. Then look into why Clay Gordon was at the hearing. What's his involvement in this mess? James, you go over the facts in the case. Go back as far as you can, look at the project's financial reports and disclosures. Both the money from the federal government and the private sector money. Have Elizabeth help you in answering any questions about private donors. I want to know where every penny came from and where it went. If you need to have any information that's not available to you, call Jake. He has a way of finding things hidden on the internet. We are looking for information and involvement by anyone or anything outside Elizabeth's normal circle of associates."

Jake quickly said, "Don't use your phones to discuss this case. Especially never inside your homes, office or car."

The group looked at him like he was crazy. "You think I'm nuts and paranoid about phones and security. But, I know how easy they are to hack into your devices and track your whereabouts'. Give me your phones and I will install an app so we can communicate about the case. When you send a message using this app it will erase itself as soon as the person you sent it to has read it. Prevents anyone looking into your phone and finding messages. At least I hope they cannot. Works for kids and drug dealers, I'd guessing it will work for us. Humor me with this. I spent a lot of years doing research into cases and the information leaks. Phone bugs and cell phone taps are the easiest ways to find out information. Trust me, I used those same methods myself. Sorry, I should not have said that. Please disregard my comments."

James looked at Charlie with a serious look. "Are you saying what Jake is doing is illegal? I don't want my career started off with charges of illegally obtaining evidence. My career could be the shortest in history if you take me down that road."

Charlie chuckled as he patted James on the shoulder, "My boy, I am not sure what Jake does, but he will do it in the most ethical way possible and if he gets caught, we'll give him the best representation we can. Don't worry, he works for me. If the evidence is clean and above board, it will be available to us. But if for some reason there is something illegal, those parties who have a hand in it will not want attention brought to it either. Sometimes all you have to do is shake the cage of those involved in a conspiracy and they begin to act suspicious. Embezzlers are good at hiding money and working in the shadows of the financial world. They are not always good at pretending to be innocent or covering their tracks, especially if they are not in it alone in their crime. Everyone is always worried about the other person making a mistake. I don't think we have that in this case. Besides, Jake's skills in unearthing information in a case often hinge on the fact people want to text or communicate to reassure themselves they are safe. People can't keep their mouths shut or their fingers off their communication devices. They are often non-professional criminals and they will want to save their own skin and not go to jail while ratting out someone else to take the blame. We just have to find a nervous rat who likes to text messages if he thinks we suspect something. And Jake is good at finding nervous rats. We have to be careful not to overlook information about the ADAM Project. If in any of your investigations you see a reference to the project, red flag it and let Jake know. Under no circumstances will Elizabeth or James go after evidence out in the real world. If this is about Elizabeth's discoveries and it is worth billions, whoever is after the information may be willing to take her if they cannot get their hands on her research."

Jake chuckled as he said, "Yea, I am pretty good. That's why I worked for the government in the first place. Besides, when I find my rat they may still have coffee creamer on their shoes."

As Jake finished he turned to Elizabeth, "I hate to ask you this, but may I have your cell phone for the evening?"

Elizabeth became instantly angered at the suggestion she was guilty of a crime by those representing her. "I won't give you my phone, you have no right to go through my personal information. I have told you I don't know what's going on and that is enough. Besides, I thought you took on my case because you believed I was innocent?"

Jake smiled as he said, "No, I took on this case because Charlie needed my help and the crappie weren't biting. If I am going to do my job I need to know if I am working with a marked deck. Your phone contains a list of who you called and when. I am just going to look at who you have been communicating with. If you are being set up, good chance it is someone you know, or at least have some association with. You are too close to see patterns, I am not. I'll also install the privacy app so we can communicate with each other."

Charlie placed his hand on Elizabeth's shoulder. You called me, remember? I don't know anything about this case other than you are accused of an embezzlement. You say you are innocent of what they have charged you with. Up to that point I believe you. But there is something going on around you. I believe it is about more than the money and that scares me. Remember, you are the one facing jail time. I think you want the truth also, that's all Jake is looking for."

Elizabeth fumbled around in her purse as she retrieved her personal phone. After a while she looked up at Jake as she offered it to him. "I don't know what you think you are going to find, but I know you'll be disappointed in the information in the phone."

Jake carefully took the phone from her hand. "I am sure you are right, but at least I am asking. The government may have already tapped your phone. If they think they have probable cause during the investigation, they would have gotten a warrant to tap your phones. I want to know as much as they do. I will get it back to you tomorrow. Under no circumstance should you contact me or Charlie on our everyday phones, only use the app I'll install on your phone. Trust me, at least for a while we can communicate with each other and not have to worry about someone listening in on our conversations."

Downing broke the tension of the moment as he said to Jake, I'll install the app, which one are you using?"

Jake swiped his phone and showed Downing the app, it's called, Invisible Ink. Kind of like when you were a kid and wrote a message in lemon juice to make your words invisible. Same thing. Only once you open and swipe the text, it is gone just as fast. So don't mess around when you open the text. Read it quick because it's going away forever.

Downing then turned to Charlie, "What are you going to do next?"

"I have someone I want to talk with. I need to try and understand more about genetics. I can see this is more than the money. I got myself into this, so I need to have a good understanding of the conversation… if it becomes part of the trial."

Elizabeth quickly replied, "I can tell you all you want to know about the science of genetics. Just ask the questions."

Downing leaned forward toward Charlie, "I would like to know more myself. Who are you going to talk with?"

Charlie took a drink of his coffee. "I appreciate your offer Elizabeth, but I need someone not involved in the case to speak with. No personal agenda, just answers about my questions."

Elizabeth sat back in her chair showing a bit of disgust with Charlie's answer.

Charlie then turned to Downing, "You concentrate on the money and charges the government has brought. Jake will help you and so will Elizabeth. When I get back I want you to know everything possible about the project finances and how the money was used in the project. Jake knows the process. Just plain old getting into the paperwork and asking questions. Don't be afraid to speak with the university's controllers for the program. When we get into the courtroom I want you to handle the embezzlement charges. And if…"

He paused as he thought for a moment. "I don't think Harrison will want to go into the science and the implications of the research. He's no expert and the possibilities of the case slipping away because of his not being prepared for the science, he's not likely to go down that path. He reminded me of his long term goals. A trial on science and its implications for mankind is a constitutional loser, even if he wins. Not something he wants to be part of."

Downing stood and said, "what about my office? If we can't talk without being afraid someone is listing, where do I go to work?"

Jake popped up quickly, "I got that covered, I need to make a call to a friend who is retired from the FBI. Does security sweeps for companies. He doesn't ask questions, just finds the bugs. He can also install a device that will detect bugs in the future. By tomorrow your office should be cleared, if anything is there."

With that the group left the coffee shop and returned to the office. Jake cautioned them to get what they needed from the office and leave without saying anything about the case. Charlie told Downing to make sure he checked the file cabinet and try to determine the files that were disturbed or missing.

With those final instructions the group entered the office. Charlie stopped Jake as he headed up the steps. "I want you to keep an eye on things for the few days. Call me on the burner or send a text on the spy app you put on my phone. Let me know if any files are missing."

Jake looked a Charlie and shook his head, indicating he understood his request as he asked, "Where you going?"

"I'm going to England to visit your friend at Stonemore. Let him know I am coming and give him some idea of what we are into. Don't tell Elizabeth or Downing where I am going. I hope I can get some direct answers to my question about the science in this case. I don't like being in the dark about anything that could come up in the trial. I have to understand the conversation about all this genetic stuff. And somehow I know Elizabeth is not being completely truthful. Something is missing, I hope your friend can tell me what I need to know."

Chapter 15

The Secret of Light

Charlie was standing on the edge of the creek in the swift water. He watched as a dozen gold eyed Mayflies floated by from an upstream hatch. He knew this was the morning. The morning when he would lure The General from his hidden place on the far side of the creek.

Carefully he rolled his line to a spot just above the hiding place of his quarry. His cast looked so natural, he just knew The General could not resist the temptation of his offering. This was the day he would finally catch his old nemesis, "The General".

After a dozen casts he saw a dark shadow move below his fly. Suddenly the water erupted as the fish took the fly as it leaped from the cold waters surface. Charlie's eyes lit up as he said to himself, "IT'S HIM… By damned, I've finally got him."

For the next few minutes he carefully walked downstream, being cautious to keep the fish's head high. He knew this would tire the fish as he struggled. Finally as he maneuvered him into a quiet part of the stream, the fish's movements became more slowed and he rolled onto his side. Charlie had worn him down with his angling skills. Now, all he had to do, bring him to the surface and carefully pick him up. Charlie's mind raced ahead. "What am I going to do with him when I catch him? I never let myself think about that. It's always been about me vs. him in the canyon in the dark swift waters. Now, I have him in my control. I have won, but that victory came at the cost of this fish's freedom that I have long sought to catch. I have won this time, but how many times had he out smarted me." Somehow the victory was not what Charlie had dreamed of.

Finally the fish gave up all resistance. Charlie slowly pulled the magnificent fish to within his grasp. "If I can just touch the fish, I will have won."

The fish was now just below Charlie as he held his pole high, to keep the line tight. He could see the fish gasping for oxygen in the cold water. As he opened his mouth Charlie reached down and grasped the fish by his lower lip. He was being careful not to injure his old friend. At that moment he felt like all was right with the world.

But, at that moment he saw a flash of something coming from below the darkness of the deep water. It was big and coming unbelievably fast. He quickly pulled The General from the water. But just as he lifted his prize from the water's surface, a flash of gold reflected in the morning sun. It was enormous, nearly four feet long. It was the biggest Golden Trout he had ever seen. And the gigantic fish had a mouth full of razor sharp teeth. The monster Golden Trout took half The General's body into its mouth. It slammed its powerful jaws shut, shearing the once beautiful fish in half. This happened right before Charlie's eyes as he held his longtime prize. The golden trout fell back into the water and disappeared, leaving only a trail of blood on the water's surface. Charlie was stunned as he watched The General's bright red blood run down its torn gut and stain the clear cold water. He looked at the fish's face. He knew it was dead, but still the fish's mouth moved, gasping for air. Its eyes did not show pain or sadness. It only stared at Charlie, wondering what he done so wrong to die like this. Perhaps it realized Charlie had not done this to him. After all, he and his old adversary had a bond of trust in this ancient canyon, far from the world of man.

Charlie looked up at the blue sky and screamed. "NO…GOD…NO!!!! He had not felt this pain since his family was killed.

In the fog of his pain he began to hear a voice, "Sir… Sir, are you alright?"

Charlie suddenly awakened and realized he had been sleeping. The death of The General was just a bad dream. He looked around and could see people staring at him. Again the red hair stewardess asked, "Sir, are you all right. You seemed to be in some distress as you slept. May I get you something?"

Charlie gathered his senses and said, "NO… I mean yes, I just had a bad dream. Too many American horror movies I guess." With that he gave her a warm smile, which she returned.

"Mr. Morgan, we will be landing at Heathrow in twenty minutes, can I get you anything?"

Charlie adjusted himself in his seat and said to the red haired lady, "A cold glass of water please." As Charlie waited for his water he looked out the window onto the Atlantic below. He could see a cargo ship moving along toward England. He strained his neck to see as far forward as he could. He began to see the outline of land in the distance. Charlie remembered it had been a long time since he was in the British Isles. He had been stationed in Germany when he was in the service. Had an extended layover one time heading home. That was a long time ago.

"Mr. Morgan, your water." He thanked the lady for her kindness. He now felt a little embarrassed as to her having to wake him from his terrible dream. He thought to himself. "That bastard golden trout haunts me in real life and now it is terrorizing me in my dreams. It's time to take up some sport where the object of the sport doesn't bite things in half."

An hour later Charlie was making his way through the busy terminal. He was looking for some sign of the man he was to contact. Finally after five minutes he saw a young man holding a sign. He was much too young to be the man Jake had told him about. But still, he had a sign with his name on it. "STONEMORE… CHARLIE MORGAN."

"Young man, that's me. Are you, by chance, Angus Whitechurch?"

"No sir, my name is Robert Pincer. I am a student of the Professor Whitechurch. He has a meeting and he asked me to fetch you from the airport. Do you have any other luggage?"

Charlie shook his head no. "Can we go now? It's been a long trip and I want to stop by my hotel and clean up. Are you in a big hurry to get me back to the university?"

"No sir, I believe your hotel is on the way. Won't be a problem."

Charlie started to get in the car and noticed the steering wheel on the wrong side. He stopped abruptly, "Sorry, I forgot you all drive from the wrong side."

The young man said in response, "You will excuse me sir, but I believe it is you Yanks that have everything backwards." He laughed.

The sun was warm on Charlie's face as they made their way through the busy streets. Before he knew it he had fallen asleep once again.

He was awakened by someone touching his shoulder. "Sir, we are here… at your hotel. I'll let you out and park the car. I'll meet you in the lobby."

Charlie realized he was more tired than he had thought. As he was checking in the clerk said, "Mr. Morgan, you have a message." With that he handed Charlie a piece of paper. It was from Whitechurch. Forgive me Mr. Morgan, but pressing business has come up at the university. I have made reservations at the Parlor House Restaurant for six-thirty. I will see you then."

Charlie looked at his watch and realized it was still on US time. He looked around the room until he saw a clock near the elevator. Two-twenty five. He realized it was several hours until the meeting and he could get some sleep.

About then he saw the young man coming in the door. "Mr. Pincer, Whitechurch has postponed our meeting until six-thirty this evening at the Parlor House. Is that near or far from here?"

"Not far sir, it's just a couple of streets over. Can't miss it, nice place. I've never been able to afford it, but some of the fellows say it is nice. If you like, I can come back this evening and take you over?"

"No, thank you, I think I'll get a couple hours sleep, let my body catch up on the time change and walk over. Never been in this part of the city before. Like to do a little sightseeing. Is it by chance a part of town you would not want to be alone at night in?"

Pincer answered, "Lot of university folks and tourists stay in this part of town. Well-lit and pretty safe. You'll be fine. But, let me give you my number. If you would need anything. I'll see if I can help."

With that the two shook hands as Charlie thanked him. Charlie opened the door to his room. The curtains were closed and the room was dark. Good, he thought to himself as he laid down on the bed. He remembered he had been awakened twice today already. He had better call the desk and make sure he did not oversleep this evening's meeting. It wasn't long until he was asleep.

The sound of the phone broke the silence in the darkened room. Charlie fumbled around until he lifted it and then replaced it without saying anything. He pressed the light on his phone to look at the nearby clock, it showed five. He had enough time to shower and awaken from his tiring day.

It took Charlie a half hour to ready himself and make his way toward the hotel's elevator. As the doors opened, he saw the young man that had brought him from the airport waiting in the lobby.

"Robert, wasn't it?"

The young man smiled as he said, "Very good, for a Yank.

Charlie half chuckled as he motioned the young man towards the front doors. In a few minutes the two men were walking along the street.

"Are you an assistant to Professor Whitechurch?"

"Yes and no. I am a grad student working on my advanced theories of light properties when exposed to gaseous elements. After I finished my studies I was asked by the professor to stay on for a year. He wanted me to help tutor new students. In return I worked in his labs on my theories. Worked out well for the university and for me, at least for a while. What about you, are you a scientist?"

Charlie smiled as he looked at the young man, "No, I am an attorney. I wanted to gain some understanding of genetics and science before I go into the courtroom. Not likely I'll have to know a lot about the subject, but at least I want to understand the terminology. I'd rather be ready if a scientific word is spoken. Having some understanding and not have to say, "Dah, what's that mean?" Robert laughed at his comment.

As they walked, Robert pointed out the many fascinating places in Stonemore. It is a town where higher learning has been around a very long time. "Simply to be able to say you attended Stonemore University is a feather in a person's cap."

Charlie looked at Robert and said. "Very nice indeed. But do you have a football team? In America we have lots of great football teams, does Stonemore have a football team?"

"We have a rugby team, something you Yanks don't play. I think it is a bit too rough a sport for you all. Does that count?"

Charlie looked at him and said, "In the spirit of diplomacy, I'll say yes. But, if you are ever in America, give me a call, I'll take you to a football game." The men agreed on the deal.

Before long Robert pointed toward an older looking English building on a narrow street. "There, over there. That's the Parlor

House. Fine place, I believe your host will be along shortly. He said go along inside and he would catch up with you. They know you are coming. Enjoy your meal."

Charlie looked a bit surprised, "You're not coming in with me?"

Robert paused as he stepped closer before answering, "No, I have some work to complete back at the lab. "Enjoy your meal and I may see you at the university if you come by."

Charlie and the young man shook hands. In a few moments Charlie walked into the restaurant. A very tall and formal looking man asked, "Sir, do you have a reservation?"

Charlie stepped closer as he said, "No, I don't, but Angus Whitechurch should."

The tall man arched his eyebrow as he said, "Very good sir, would you please follow me. Doctor Whitechurch asked for his special table. Close to the front window and in an uncrowded area of the restaurant."

The man directed Charlie toward the stairs that went up to another level. The second floor was away from the noise of the lower dining area. Toward the front overlooking the street was an area that resembled an indoor garden with lots of natural light. Surrounding the room were a multitude of plants and small trees. The air had a hint of jasmine which grew near the far wall. After a few moments the tall man pointed towards the table next to a small decorative waterfalls.

'Sir, may I get you something to drink while you are waiting for your host?"

"I would like a good English stout please. Been a long time since I've had one."

"Very good, sir." With that the tall man disappeared toward the stairs.

Charlie looked out at the street below. As the evening came on, it was becoming less and less busy. The streets seemed narrow by American standards. But, he realized that England's history goes back far beyond that of America's.

From behind him a voice said. "I take it you are Charlie Morgan?"

Charlie turned and looked up at two men. One man was of average height and looked to be in his late fifties. He wore the collar of a minister or priest. The other man was older, with a bit of a paunch stomach. As he removed his jacket, his red suspenders showed the strain of holding his pants in position. He looked to be in his late sixties.

"Yes, I am Charlie Morgan. I am assuming one of you gentleman is Angus Whitechurch?"

The man with the red suspenders offered his hand and said, "I'm Whitechurch and my friend is Tucker Hunter. I just call him Tuck. He's a friend and keeps me on the straight and narrow. He's a little stiff sometimes, so don't pay him no never mind."

He leaned closer to Charlie as he said, "I think his clerical collar gets a little too tight sometimes."

Tucker smiled and offer his hand to Charlie. "Hello, my old friend here seems to think my collar gives me some influence with the almighty. Even if it did, it is unlikely I could help Angus with his transgressions against several of the Ten Commandments."

The men shook hands and sat down. Angus asked, "How's Jake these days? I haven't seen him in several years. I use to enjoy talking with him about the cases he was working on. And speaking of use to work on, I thought after his wife died he went into self-exile on some lake."

Charlie responded, "Well, it's true he spends a lot of time at the cabin, but this case came up and I needed his research skills on the embezzlement part of the case. Jake was one of the best when we worked for the Justice Department. I asked him to come out of retirement and help out with this case. To be honest, he put up a little struggle, but I think he was ready to get away from the lake for a while and return to work. At least for a while. He's still good…and just a little more paranoid then he used to be. If you are wondering why I came instead of Jake, well I'll be helping to handle the case in the courtroom. And I have some questions of my own I need to have you answer. I am hoping you can give me a crash course in genetics. I know I cannot fully understand the technology, but I need to be able to understand the conversation."

About then the tall waiter returned and sat Charlie's drink on the table. He looked at Angus and Tucker, "Gentleman, can I get you something?"

Whitechurch pointed towards Charlie's drink," I'll have one of those and bring me a shot of whiskey, make that double."

Tucker looked at Charlie and said, "You see what I mean about keeping him out of trouble. He then looked at the waiter, "I'll take an Irish whiskey, and make it a double."

Angus laughed, "Take care of me you old reprobate. You just take care of yourself." They both laughed.

Angus then turned his attention to Charlie, "This place has the best steak in this part of the country and you Yanks like your meat red and rare, don't you?"

Charlie said with a bit of a smile, "Steak sounds good, and yes we like beef if that's what you are referring to as red meat."

Angus looked at the waiter and without asking the others said, "Go ahead and put in our order for three medium well steaks. In the meantime we'll drink this fine Irish whiskey and get acquainted." The waiter tipped his head as he turned to leave.

"Well, Charlie, let's get down to business. I understand you and Jake are working on the case your government has against Elizabeth Mathenson?"

"Yes, that's right. She's been charged with embezzlements of funds from her program. They have pulled her tenure and dismissed her from the university. Now the government is charging her with misappropriations of funds. They are casting a wide net of accusations in her case."

Whitechurch leaned forward quickly and bellowed, "Pull her tenure, them bastards can't do that!"

"I know they can't, but they sure as to hell have said they are going to. This case has a lot of twists and turns. In fact when I first got involved, I was given to believe it was a simple fraud case of government grant monies. It is anything but simple."

Angus paused before asking, "So what do you want from me. I don't know shit about American law and embezzlement."

About that time the waiter returned and sat their drinks on the table. "Will there be anything else at this time gentleman?"

Whitechurch, without looking at the man, said, "No…not now. But, bring me another round when the steaks come. Charlie, you want anything else?" Charlie smiled as he looked at the waiter and shook his head, no.

Whitechurch drank his drink straight down without stopping. As he sat the empty glass on the table he looked at Charlie saying, "Well?"

Charlie began to explain. "When I first took this case, I was led to believe it was about money, but as time passed, things came to light and I began to believe there was more than just money involved. I've began to think the case may be about the doctor's research; the money is just a distraction. Someone may be using it to push her to do something she's not willing to do. And to tell you the truth, I do not like not knowing the full truth in a case I am working on, especially when I am getting ready to go into a court of law to defend my client. Jake thinks you can help answer some questions about the research and what Elizabeth has gotten herself into."

Whitechurch sat back in his chair and made himself comfortable. "I don't know what you know about her work, but I'll tell you what I know. At least my part in all of this. What has she told you so far?"

"At first I was thinking only about the embezzlement charges. Her lab work was not even a consideration for me. She's told me about her research, but I am thinking she told me only bits and pieces. That makes it impossible for me to understand. Especially when trying to figure how the federal charges and her research fit together. I know she does genetic research, something she has spent her life working on. Early in her career she had some insight into the possibilities of biogenetic medicine and the potentials it had. Her goals, she explained to me, were to find cures for disease by modifying DNA. But, I don't understand cloning mammoths and her trips to Israel and Scotland and here to England in all of this. I know somehow it all fits together, but I can't put it straight in my thinking. I am a lawyer who spent most of his life working with fraud cases. Evidence, the law and finding justice. This science is outside my expertise. I need to understand the conversation in the case. If I don't, she's going to jail for a very long time. By her actions I suspect she is guilty of something, but I don't think it is embezzlement."

Whitechurch took another drink and emptied his glass. As he sat it on the table he looked for the waiter wanting to have his glass refilled.

Whitechurch began, "Let me try and put what I know into context for my part. Elizabeth's team for the last ten years has made some great discoveries with DNA. Things just a few years ago that were believed to be far outside the reach of science, suddenly they weren't. Do you remember when they grew what looked to be a human ear on the back of a mouse?" Charlie shook his head he did.

"Well, in a way that's when it all started. The ear and the mouse wasn't what people believed it was. People started saying terrible things about genetically created parts of humans. But, even though the science was not about genetics it did give scientists an idea about building a biodegradable framework and inserting cells. Under the right circumstance they did grow. Then Elizabeth discovered she might be able to construct a matrix frame for a synthetic DNA molecule. Seed it with protein fragments from real DNA genes. That's what they did with the mouse's ear, only on a much larger scale. Trouble was, technology and the science was not there yet. And her program had ethical boundaries she couldn't cross. She knew that scientists in Korea were creating protein structures from original parts of the body and placing those cells into the matrix to grow. For some reason they could not get the matrix to accept the genetic material and grow properly. Best they could do was to recreate an ear on the back of a mouse by placing living cells from a human ear into a matrix and giving it a place to grow. Mother Nature did the rest. It was more a side show novelty.

About the same time, Korea, with support from Russia, began a program to clone a woolly mammoth. The Russians found a well preserved mammoth in the Siberian Tundra. They joined forces with the Koreans who had the technology to clone small animals. At first they didn't want anyone to know. Then about the same time a sheep was cloned, making the front page of world news. The Koreans were quietly cloning dogs. People with lots of money could clone the same dog over and over. All done with the modern techniques of genetic science. Anyway, when the Koreans were trying to create a new woolly mammoth from the DNA they found in the frozen carcass, they realized the DNA was badly damaged. Not all of it, but some. They came up with a way to program protein molecules to act like clippers and remove the damaged section of the DNA strain. Then they would do the same thing from a living elephant's DNA. Other proteins were programed to replace the damaged DNA matrix of the mammoth. This discovery changed everything. Science was engineering a new woolly mammoth. I believe in the very near future they will announce they have successfully repaired a mammoth's DNA and placed it into a modern elephant's egg. They will then place that egg inside a living elephant. Twenty-two months later and the world will behold a furry elephant. People will line up and pay good money to see a creature long gone from this world. It won't be real, but most won't care. In our scientific greatness we will have played God. They'll be so much interest in seeing it, no one will ask what was done to create this long dead creature. They will call it progress, but at what cost?" Angus drank his drink straight down and motioned for the waiter to bring another. Then he sat quiet staring into the empty glass.

Charlie paused for a few moments before saying, "Elizabeth has told me her work is not about cloning an elephant. Did she lie to me?"

Whitechurch saw the waiter out of the corner of his eye and once again shook the ice in the glass at him. The man acknowledged his impatient gesture and hurried off.

"No, she didn't lie to you, she's not about replicating some old dead elephant. That thing will just become a sideshow freak. But, not only a sideshow freak, it is a distraction from the scientific community's real work. There are those that figure if they proclaim they have created a living mammoth from ten thousand years ago, people will be drawn to it, along with their money. Truth is, if they could create a T-Rex they would. Not for science, but for profit. Unfortunately behind the real science there are scientists who are working to do the ultimate recreation, mankind. I have to tell you that I believe that is what Elizabeth has gotten herself into. Not for sideshow reason, but for reason that will be the betterment of mankind. Cure diseases and damaged body parts. Her intentions are true. But, there are some that believe science is not suppose to go there. Just think Charlie, the minute they resurrect a mammoth from dead DNA, someone is going to ask, can you bring back my dead wife or father or mother. And that person has the money to support that question. That day is not far off."

About then the waiter returned with two glasses of dark stout along with two whiskeys. Whitechurch picked up the whiskey and drank it down once again without stopping. When he finished he sat it hard onto the table.

"Where were we, OH YES! Elizabeth attended a seminar in Israel last year. They don't want the world to know, but they have an extensive research project dedicated to tracing people's DNA back to its origin. They are working to prove scientifically they came from the same region of the world as their ancestors were, as written in the Torah. You know exodus from Egypt and slavery, wandering in the desert, all that history. They want to prove they are the chosen people with real science. Anyway, Elizabeth talked, well I won't say talked she negotiated to purchase ten thousand duplicate DNA samples. The Israelis took multiple samples and were more than willing for a price to sell them to her. She in turn sent them to my lab to be cataloged. She had always believed DNA could tell us more than just the origin of the person. She believed we could analyze each sample with pulse diode technology. Having so many from the same region of Israel, she hoped to build a complete DNA blueprint. I don't mean just analyze, I mean look into each chemical element and its precise position in the elemental structure arrangement. Then use that blueprint to program proteins to reconstruct DNA, element by element. She believed that created DNA would be perfect. Just as it was in Adam. By the way this is the point where she came to me for help, I'll show you tomorrow what she needed to continue her work. Anyway, in her research she found genes with abnormalities and damaged sections. But, then she remembered the mammoth DNA and how they programed protein molecules to remove and replace undamaged genes from a healthy modern elephant. With ten thousand DNA samples analyzed she planned to map the DNA structure of mankind. That map would be from the supposed birth place of what we know today…"

Angus seemed to become distracted from his explanation as he called out. "Where's our steaks? I'm hungry, damn it!" Angus took another drink looking for the waiter.

Charlie sat for a moment not saying a word as he could tell Angus was feeling his whiskey. He then looked at Tucker, "You have been quiet during his explanation. How does this all sit with the church?"

Tucker grinned, "Well Mr. Morgan, you are not shy about the tough questions. On the one hand it sounds to me like Elizabeth is playing God. On the other hand it sounds like she is using science to prove that the Bible is correct in the origin of man. The question is, if she can somehow do this thing; is it by the hand of God or does it remove Him from the equation. Now, how does that explanation fit in your defense for her embezzlement?"

Charlie just shook his head in seemingly disbelief.

"Gentleman, your steaks are served." With that the man sat their dinner on the table asked if he could get them anything else.

Charlie shook his head indicating no. The waiter looked at Whitechurch, he was finishing his glass of whiskey. "I think one more will help me finish my steak." He smiled at the waiter.

For the next half hour the men spoke of family, music and even a little politics. Charlie could tell Whitechurch had drank a little too much as his speech was slurred just a bit. But, still he was a man who could hold his drinks, up to a point. He finished his meal and thanked the waiter for his outstanding steak.

"Mr. Morgan… Charlie…Before I finish my story I would like for you to come by the university tomorrow. You remember I told you Elizabeth came to me for help with a problem."

He looked around the room and said as if he were whispering, "Shhhhh… this is a secret, but I want to show you the secret of light in my lab tomorrow. With the explanation I have given you tonight, and what I will show you tomorrow, you will have an understanding of the world in which Elizabeth works. I don't have any reason to believe her intentions are anything but good, but I believe she has gotten lost in her pursuit of science and may have gone into a dark place looking for the ultimate answers. Come by tomorrow, Robert will fetch you to my lab, eight o'clock sharp." He stood and held onto the table to balance himself and then looked at Charlie with an arched brow, "Make that ten o'clock.

Charlie and Tucker helped him down the stairs. Charlie was not surprised to see Robert waiting in his car on the curb of the narrow street.

The young man stepped out and said, "Sir, did they serve you too much bad whiskey again?"

Angus looked at Charlie, "See, I have a very well trained assistant. See you in the morning." With that he sat down in the car.

Tucker offered his hand to Charlie, "Mr. Morgan, it has been a pleasure and an enlightened conversation. I have a bit of advice for you. I am guessing you are not a man of God. Nor do I believe you are a total non-believer, just someone who has never taken the time to find the truth. Something for you to think about. Scientists searching the Arabian Sea have found a piece of wood in the mud bottom. They have declared it is a piece of a tree. They have determined it has been down in the bottom mud for a very, very long time. In fact using modern radio carbon dating techniques, they place the age of the wood around six thousand years old. Given the age and location of the find, they believe it is part of a tree that once grew in the Garden of Eden as stated in the book of Genesis. Scholars will for many years debate whether the piece of wood is or is not. Those who believe in God will take it as the truth written in the Bible. For others it is a muddy piece of wood. Being that I am a man of God, there are those who say I should believe in blind fate and proclaim Eden has been found. But, you know what Charlie, God gave me the ability to think. Above all other living creatures, God in his wisdom gave me the greatest of gifts, intelligence. I don't think He would have done that if He wanted me to believe blindly. My advice to you in this case, open your eyes and your mind. But, mostly open your heart to the possibility of life, no matter how it begins. I'm guessing even Darwin believed all life is a miracle. It is only man that questions the origin of that miracle."

With good-byes exchanged, Robert closed the door and the car moved away. Charlie looked across the street and saw an old English style shop. In the window was a sign neatly printed, Fine Hand Rolled Cigars. He decided to indulge his old habit as he walked back to his hotel on this pleasant evening. He enjoyed the walk and the taste of a good cigar. He stopped on a bridge and looked down at the stream below. He wondered if somewhere in this stream, was there a fish with a name given to him by a fisherman. He realized how funny that must have sounded.

Chapter 16
Inside the Light

Charlie was still tired when the phone rang awakening him. In the back of his mind he knew his internal body clock was not matching the world around him. He realized a hot shower and coffee would help to get him going.

An hour later he was sitting in the dining room drinking coffee and eating a bowl of tasteless oatmeal. As he looked toward the door he saw young Robert enter the room.

"Good Morning Mr. Morgan, I hope you slept well? It is going to be a fine day."

Charlie motioned with his hand for the young man to sit down in the chair across from him. "Robert, I will have to say you seem to be everywhere. Surely you must be more than a grad student working for Whitechurch?"

"I don't look at it like work. I have been given an opportunity to study with a brilliant man whose work on light waves is the basis for several great scientific advancements. I could have moved on and gotten a good job in the private sector making decent money. Even had an opportunity to go to America to work for NASA. I could have been part of the team building the James Webb Space Telescope. Instead, I chose to stay here. I can learn more in a few years about light spectrum analytics than I could in decades at NASA. Angus Whitechurch has the ability and understanding to see the structure of light and its secrets. He is getting on in years and, as you saw last night, he sometimes drinks a little too much. Still he is a great man of science. He has taught me a great deal and if he needs me to help him get home on occasions, that's okay with me."

Charlie signed the check and drank the last of his coffee. "Well Mr. Pincer, are you ready to take me to Stonemore?" The young man smiled and indicated he was.

It didn't take long until they arrived in front of an ivy covered building. A great many people were coming and going around the building. Robert exited the vehicle and headed inside the stately building. Once inside he pressed the button calling the elevator. Almost instantly the doors opened and the two men went inside. Robert pressed four.

The doors opened into a hallway with high arched ceilings. It was an old building, but well kept. Charlie's mind briefly knew these hallways have been the home to education of the highest level for hundreds of years.

Halfway down the hallway Robert pointed toward a door. The sign outside read, Angus Whitechurch. Charlie open the door slowly and peered inside.

"Come in, come in Charlie." He pointed toward Tucker Hunter who was already drinking a cup of tea. "Charlie, you remember my friend Tuck from last night. I hope you don't mind if he joins us today. He understands everything that science sometimes doesn't. He pretends God gives him the answers."

Tucker stood and offered his hand. "Charlie, we don't stand on ceremony here. Can I get you a cup of coffee or even better a good cup of English tea?"

Charlie responded, "No thanks, I just finished breakfast and my morning cup of coffee. Before we begin, I just want to thank both of you again for taking the time to help me with my understanding of the science in this case. Tucker, you may be able to help with the religious implications, which I don't fully understand either. Hopefully the case will stay on track to be only about the money, easy and understandable. But you gentlemen's insight into the science and religious aspect will help me understand the scientific terms if they come up in court, but that is highly unlikely at this point."

Angus sat back down as he said, "Well, Charlie, I bet you don't know, I have doctorates in both astrophysics and medicine. As a sideline hobby I became an ordained vicar hoping to find answers in my soul. No luck. The real truth is, I am more of an engineer than anything else. Never been married and Stonemore is my home and family. I have always loved to study and discover new things. I guess that's how I ended up being so well educated, and a poor excuse for a human being. I have always looked inside of things to see how they work and what they are made of. Space, the human body and even my soul, that's where the vicar part comes in. I've had some success with interpreting what I find in space and analyzing it. I even had some success in understanding the mechanics of the human body, especially DNA and its makeup. I would have to say to date, my work with analyzing light from the stars has given other scientists new opportunities. My studies have helped analyze chemical signatures by understanding the spectrum of light. My department's good fortune in guiding that research has benefited everyone. Did young Robert tell you NASA wanted him to be part of its new telescope team? They knew the best and brightest could help with their next space telescope. And truth be told, he would have done well to go and be part of it, but for some reason he chose to stay and help me. And as you know from last night, I am grateful for his help. Oh yea, one more thing that I am, a drunk."

About then Robert stepped into the room, "Sir, I have things prepared in the lab. You gentleman should go into the dark room to prepare your eyes."

Charlie looked at Angus, "Prepare our eyes?"

"Yes, we will be going in the light spectrum studies lab shortly. It is kept dark to try and prevent contamination from outside light while the equipment is in use. There is a small room next to the lab where almost no light can enter. Just enough so you won't fall. Follow me please."

It did not take long for the men to make their way into the dark room. Inside there were several chairs around the outside wall. Overhead a small bulb illuminated with just enough light that Charlie could see to sit down.

Angus sat down across from Charlie. In a few moments Angus said, "Our eyes need to adjust for about ten minutes. You will begin to notice the longer we sit here the brighter the room will become. It really doesn't. Your eyes adjust to the darkness allowing in more light. While we wait, it will give us a chance to talk. I take it Doctor Mathenson is in a lot of trouble. More trouble than just the money she is accused of taking. And I assume you know about the DNA samples the Israelis are now demanding to be returned. If you don't know, she got them legitimately. She saw an opportunity to obtain a large data base of DNA samples. For a price, she obtained them. Sometime later, an Ultra-Orthodox Rabbi found out about the deal. He became enraged that a non-Jewish doctor had ownership of the DNA samples. A non-believer had ownership of the very soul of the Jewish people. Well, that Rabbi had several friends in the Knesset. They put pressure on your State Department to return the samples. Elizabeth refused. They are still trying to force her to give up the DNA samples and return them to Israel. Now, I'll tell you what you don't know. Before I do, I ask that you never reveal to anyone what I am about to tell you. Those samples have already underwent pulse diode light analysis here in my labs."

Charlie, could begin to see Angus's face more clearly. "You have them? But the Israelis and my government are looking for them. Computer files and those samples along with a cryogenic chamber were items mentioned in the warrant they served on Elizabeth."

"Charlie, you have to understand, this all began before the Rabbi found out about the missing DNA samples. When Elizabeth came here, there were no questions about their ownership. Anyway, back to the testing, as you know light passing through elements can be analyzed for its chemical content. My research is mainly about light from stars far out in the universe. We also test analytical equipment used in highly specialized industrial applications. My department is designing analytical equipment used to detect trace elements of explosives or biological agents in the air. Our new pulse diode technology is far more sensitive than today's technology. Anywhere trace elements can pass through a light beam, we can analyze its content. Somehow, Elizabeth got it into her head that if she could extract DNA and use this technology, we could code every element in the gene's sequence. And do it in a fraction of the time. Using pulse light emitting diodes was the key to most of the problems for both her and me. Robert, the young man who is my unofficial caretaker, he's the one who came up with the idea. Once each of the DNA sequences was coded and analyzed, and by the way we did all ten thousand of them, they were fed into our computer used to detect light sources in the sky. Each sequence was broken down into the most basic elements. And I do mean, the most basic arrangement of atoms in each element. Thirty million genes and three billion coded elements in the sequence. Not long ago it would have been impossible to work with such large numbers. After we have the data we converted it into useful information. Before we had this new technique along with superfast computer system, this process would have taken years to compare ten thousand DNA samples. By using this technique the entire group was completed in months."

About then a door opened into the light lab. Robert said, "Gentlemen, you can go in now."

As Charlie stepped into the room he was handed a pair of heavy eye glasses. On his left was a wall with very large computer screens. The six screens were arranged in two rows of three. In front of them was what appeared to be a control console. Two technicians were making adjustments. Nearby in the wall to the right of the screens there were two large glass windows. The room on the other side was completely blacked out. No light gave away what was in the room.

Robert stepped next to Charlie, "Let me explain what you are seeing. He first walked over to the console in front of the screens. "Those glasses you are wearing are designed to give you the perception of multiple dimensions when looking at the screens images. Not quite what you think of as 3-D, like in a movie theater, but close. Those technicians are controlling two robotic arms inside the darkened room behind the glass. They are programed to do their tasks in total darkness. That way, none of the samples are contaminated with outside light. The arms are programed to prepare the samples. Next, they are placed into a chamber of an electron microscope. This is where the only light they are exposed to is a pulse diode light. It is exposed to microbursts of light. In fact it is only one ten thousandth of a second burst. It is so quick, it is undetectable to the human eye. The only way we know it fired is we see a spike in light energy on the screen." He pointed toward a digital graft.

He then pointed toward another chart. "As you see here, when we fire the light, it emits an energy signature. That energy from other light sources can produce damaging heat. Using pulse diodes, prevents heat from the light source damaging delicate DNA samples. Protons pass through the DNA's genetic material, then are captured by a target sensor. We use a four-element sensor to capture the image. Normal light has three light spectrums. The fourth produced by a modulated diode is pure protons of variable light within themselves. By adjusting the energy source value we can map the genes inside each chromosome. After each and every chromosome in the sequence is analyzed. The chemical signature of every base element is cataloged in a detailed pattern. This new process can do in a few hours what used to take days, even months. A good example of how this works from a data collection standpoint. In photography there are several formats for storing digital images. For scientific standard we often use a Tiff file format. The file records each and every pixel for color, contrast and position in the image. A simple eight by ten picture can have millions of pixels of information preserving the image. This precise information gives us the ability to reproduce the image in exact detail over and over, using the original image data. Now in comparison, if we simply wanted to preserve that image for everyday use and not use a large amount of storage capacity, we could use a file format known as, jpeg. Let me give you a real basic example of what I am getting at. A Tiff file would record a single pixel as blue. If the same picture were saved as a jpeg file, a compressed graphic format, it would lose much of its detail. To the naked eye if you printed this picture, it may be difficult to see a big difference in image quality, but if you started analyzing it up close, you would see pixels with fuzzy imagery. The Tiff file would be sharp and detailed. This equipment can do the same thing, very detailed analysis of every molecule in the DNA structure. Why so detailed you may wonder? It then gives us the ability to see the make-up of DNA's blueprint for life. We can modify it, change out sections, or use it to build a whole new DNA sequence using this information. After this scan is complete we have precise chemical compounds and their position in the molecules of the DNA spiral. We have the recipe to create synthetic DNA."

Angus pointed toward another screen. "Charlie these are what we call bases. There are four basic chemicals in DNA. Adenine, cytosine, quinine and thymine." Angus pointed toward the screen indicating each of the chemical's signatures he had told Charlie.

"These structures are held together by a sugar known as deoxyribose. The real key is the structure or matrix as I prefer to call it, it is a framework made from polymer strands holding together the chemical bases in the correct sequence. What you see here is a true biological DNA structure." The image slowly rotated on the screen. Charlie's glasses gave him the impression he could almost touch the image.

He instructed the technician at the console to bring up synthetic polymer 893 on a second screen. In a couple of seconds the image on the screen changed. It looked similar in appearance to the last image. "Charlie, this is the results of Elizabeth's project using the ADAM technology. She has created a synthetic polymer strand in her lab. It replicates a natural DNA ladder matrix."

Charlie stood in front of the screen looking at the images. He pointed toward one of the screens. "What am I looking at?"

Whitechurch tapped the technician on the shoulder, "Explain it to him in more detail. The young man changed the screen once again. "Sir, what you see now is the part of the molecule science calls, deoxyribonucleic acid, more commonly known as DNA. Inside this is the blueprint for each species. Whether it is man or a flatfish the DNA holds the key to its design. Each species passes this blueprint along to its next generation through reproduction. Inside each cell is a nucleus and inside the nucleus is the species DNA blueprint. What you see on the screen is the DNA strand. This one is unwound. We introduced an enzyme that causes the DNA to uncoil. Unwinding is part of the process it uses to replicate itself. At that moment, we pass it through the diode chamber analyzing its twenty three chromosomes. What you see near the top is a small portion of its total genes. In total there are twenty thousand. In the Human Genome there are 3 billion chemical combinations made from four bases found in DNA and RNA. They are adenine, cytosine, guanine and thymine. You will see them written as abbreviations using only the first letter for each base A, C, G, and T."

Charlie stepped away from the screen and looked into the darkened room. He turned to Angus, "I am a simple lawyer. This is way over my head. In simple terms, what does this have to do with Elizabeth's case?"

Angus responded with a chuckle, "In simple terms? Well, you now know that inside each of us there is a blueprint to who we are. Elizabeth's research working with a protein micro molecules can create complete new molecules that she calls Amino Dimatrix Atom Molecule. Think of this molecule as a piece of construction equipment. It can build things. Or should I say it can construct chemical building blocks called nucleotides. After building these, she introduces enzymes that begin to assemble them into chains. The chains begin to form basic DNA strands. Normal DNA can detect damage to it sequence. Enzymes recognize the damaged parts and begin the repair of the missing or damaged parts. You remember the mouse with the ear growing under its skin on its back? In many ways the first part of Elizabeth's ADAM protein uses that same concept to build a polymer ladder matrix. In a sense, a skeletal framework of the DNA molecules. The ADAM molecule recognizes its task and begins the construction, filling in the different parts of the DNA synthetic polymer strands with the chemical bases. Enzymes detect the missing parts, and begins construction. In theory, given enough time the project can build a human DNA complete with all twenty three chromosomes. Once built, we can insert it into a human egg. The tricky part, bring that cell to life. We think a tiny electric shock can energize the cell and begin the division of the cell. With a little luck in nine months you have a child. A child not conceived by natural ways. But, a child with a DNA completely new. If the child's DNA was analyzed, it would appear to be complete and no part of the DNA would be dormant. Exactly what would that mean? I don't know and I don't think anyone else will know until it happens. One last thing, you remember the DNA samples that Elizabeth obtained from Israel? They all have been broken down and analyzed for this research. I don't know how the Israelis are going to take hearing they no longer exist in their original form. The light analyzation process that Robert told you about has reduced the original samples to computer data. Data that can be used to build new DNA from the Israel samples. Does this created DNA mean it can create the person who it came from? No, I don't think so, but that is a question down the road and Elizabeth may have those answers. That is further down the rabbit hole than we have ventured."

Charlie shook his head as he looked around the room. In a moment he found a chair and sat down.

Tucker sat down beside him and said, "Charlie, I can see you are confused by all this information. When I first heard about this, I believed it was some wild science fiction story cooked up by some grad students with a big imagination. Science is going places some of us wonder if we should go. Your friend Elizabeth has in all likelihood found the answer to not only who we are, but what we are. She may have devised a way to create synthetic DNA, but if she has, can it be long before others will do the same thing? I teach my students, all the answers are contained in the Bible. But, at the same time I see this scientific evidence and I have to question myself, but not my beliefs. I believe in my faith. I cannot get discouraged because I see all this. After all…I believe God is responsible for everything. So He must have a hand in this science. Because I don't understand it, doesn't mean it is not real and in his control. There are many things in life He has given me no answers to, it is up to me to have faith in His teachings. Because answers have not been revealed to us, doesn't mean there are none. Perhaps God intended for us to someday return to the closed garden."

Charlie did not respond to Tucker's comment.

Angus stepped close, "Charlie is there anything else I can help you with? It gets more complex in the understanding of DNA and the part it plays in all this. But, I think you now can at least understand some of the conversations. Let's go back to my office. I have something else I want to tell you."

The three men returned to the small darkened room. As they sat the overhead light slowly grew brighter. During this time no one spoke. After five minutes their eyes had returned to almost normal. Angus opened the door and they soon were back in his office. Charlie rubbed his eyes as they adjusted to the room's lights.

Tucker asked, "You ready for that cup of tea now?"

Charlie shook his head yes, in response to his question. The three men sat down and remained quiet for a few minutes.

"Charlie, when are you returning to America?"

"I have reserved a red eye flight that leaves tonight. I need to get back to see where the rest of the team is in their investigation. I'll give Jake a call in a little while. See what he's learned about the embezzlement part of this case. Jake is the best investigator I ever worked with, even if he is a little paranoid. I know he would have liked to have made this trip, but I needed him to find out about the case."

Charlie could see Angus had something to say and it was not something he was eager to talk about.

"You said you had something else to tell me, by your hesitation I am figuring it is not good news. But, to tell you the truth I am still trying to absorb what you have shown me. So I don't think it could get any worse."

Angus leaned his elbows up on his desk as he looked Charlie in the eyes. "I don't know how far Elizabeth has taken her science, but I know she had a plan to use this new technology for several purposes. But, the one she was most excited about, is the one that scares me the most. You remember how the young man operating the equipment explained to you about the molecules and proteins working to build a synthetic DNA?"

Charlie shook his head that he remembered the young man's explanation.

"About a year ago I got a text message from Elizabeth. It was short and simple. ADAM worked and is successful. I think it is possible she's has used this technology to try and create life. When you came here, that's what I thought you were coming to find out. If I know for sure she has and how she did it? But, I now realize you had no knowledge of what she may have done. I can't say she has for sure, but I have to believe it is a real possibility."

Tucker let out a deep breath, "This institution has been around for a very long time. It prides itself on educating some of the great minds for several centuries. But, I don't think any of those that came before us ever envisioned what we believe is now possible. Wise people from all walks of life need to have the conversation of the possibilities this science raises. We need to look forward and find answers before mankind enters the creation question with a small child already in hand. It is not for us in this room to judge, but for you Charlie, you must come to terms with the possibility of what you now know is possible. I hope you find the truth and remember… I believe whatever the answers, they are at the hand of God. Only His hand is the guiding light in all creation, even where it may appear to be by the hand of man."

At that moment the door burst open, it was Robert. "Sir, you need to turn on the television. The BBC news…well, I think it is about Mr. Morgan's case in America!"

"What do you mean, my case?" Charlie asked.

Robert turned on the television and turned up the volume. The reporter was halfway through his story. "It is not clear by the reports so far, but it seems an American doctor named Elizabeth Mathenson has made a discovery of incredible possibilities. But, for some reason the university she works for has dismissed her and her government has brought charges of embezzlement against her. The case is scheduled to go to trial next month. Doctor Mathenson refused to comment on any of the information surrounding this story. To repeat this breaking story, an American doctor has for the first time reportedly created synthetic DNA and successfully inserted it into the nucleus of a living cell. Details of the process are sketchy at this time."

Angus sat back into his chair and let out a great sigh. "The cat's out of the bag for sure now. Somehow, someone has leaked to your news services what she has been working on. If I had a bottle I'd take a drink, maybe two."

He looked up at Charlie who was still considering the story. "I can't understand how this could happen. No one on Elizabeth's defense team even knew the possibility that she had gone so far."

He sat and thought for a minute before saying, "I have to get a flight back to America as soon as possible. I have no idea what is happening and I can't make any difference here." With that he stood and quickly shook the men's hands almost as if his mind was somewhere else.

Robert said, "Mr. Morgan, I'll take you back to your hotel. What time did you say your flight was to leave?"

"One-thirty. I don't think I can get an earlier flight."

As they started for the door, Tucker took a hold of Charlie's arm, "Mr. Morgan, would you mind if I accompanied you back to the states? I believe this announcement has great consequences for not only science, but for those of faith. For some reason I feel I need to be there. Maybe I can be of some help. I can gather my things and meet you at your hotel very shortly."

Charlie paused and looked at Tucker, "I am not sure why you feel the need to come all the way to America. And to be honest, I don't have time to worry about you when we get there. Something has come up with the case and I know my associates are being overwhelmed by this sudden revelation in the news. Someone is trying to sabotage this case and I have no idea why.

"Mr. Morgan, I won't be a bother. Besides, I might be able to give you some answers when it comes to your client being the creator of life."

Charlie, not wanting to delay any further, shook his head yes as he said, "It is a free country, if you want to come along, I guess that is your right. But, I don't know what you expect to find."

With that the men hurried off to prepare to journey to America.

As Charlie was riding in the car with Robert he decided to call Jake and find out what was going on. It rang only once, Jake answered, "Charlie is that you, I told you not to call me on my personal phone."

"Knock it off Jake, I don't think there could be any more secrets revealed than what I just saw on TV a short time ago. What the hell happened?"

"Well, I am not sure. We came into the office to go over the financial reports for the last ten years. Suddenly about an hour into our work, the front street filled up with news people.

Television, trucks from all the networks, reporters. They all were wanting to know the story of Elizabeth's discovery. Someone leaked information about the direction of her research and that she has created life. After that everything went crazy. About an hour later protesters showed up carrying signs about the immorality of what she has done. You remember that minister from the Church in the Sky? Well he is leading the pack, calling for the damnation of Elizabeth. Saying she has damned us in God's eyes."

The phone line broke up for a moment from interference. Then the connection returned. Jake had not known of the interference on Charlie's end. "When things got real crazy with the news people, police and the preacher's people all fighting for position out front, we slipped out the back door and made our way to that coffee shop we were at the other day. Anyway, I have moved our operations to your Elkins office. It will take the news people a few days to figure out where we are. Far as I know they don't know about your involvement in the case at this time. Charlie, you're an old friend, but what the hell have you gotten us into and when are you coming back?"

Charlie said nothing for a few seconds while he thought what to do next. Finally he said, "I'm leaving as soon as I can get a flight. Most likely I'll land at Dulles airport later this evening. I'll try and get a regional flight into Clarksburg. But, that probably won't be until morning. I'll text you when I am coming in. In the meantime try and find out where they got the story. I don't think the university or the government would have let this out. It creates as big a problem for them as it does us."

Jake asked, "Did you learn anything from Angus at Stonemore?"

"Your friend has been very helpful. I can't say I understand everything, but I can at least understand the conversation now. By the way, a colleague of Angus's is returning with me to the states. He seems to think he can be of some help. Not sure why, but he's coming with me."

"What's his association with Angus?"

"He teaches world religion. In fact he's a pastor himself. Not sure why he wants to accompany me back home, but I see no reason he shouldn't. Tell Nancy to call my housekeeper so she can prepare the spare room. I'll text you when I am sure of our flight plans. Until then, try and find out what going on with the press. And where is Elizabeth?"

"Not sure, she told Downing she was going to hole up with a friend and she would get in contact with us in a couple of days." Charlie gave Jake a few more instructions before ending the conversation.

After he finished, Robert asked, "I take it that things haven't gone well for you back home?"

Charlie sat looking out the window of the car as they moved through the busy street. Finally he answered. "I have been a lawyer close to fifty years. In all that time, I have never been associated with a case like this one. This case should have been about the money, end of story. Now, it is about so many different things I can't get my head around them all. I have to try and keep the attention on the embezzlement case. All the rest is just a distraction. If I can do that, I can get back to the quiet of my life and go fishing."

He looked at Robert, "Did I ever tell you about The General?"

Robert looked at him as he said. "You mean a general from you Yank's army?"

Charlie chuckled, "No, this general is a fish and lives under a rock in a hidden canyon, far from all the craziness of this world."

Robert got a strange look on his face as he glanced towards Charlie, "A fish?"

Abruptly he stopped in front of Charlie's hotel and then said, "You do mean a real fish that swims in a stream?"

Charlie grinned and said, "Yes, Robert, you see in America we give all our fish names. Makes it easier to identify them when we catch them from cold water creeks."

Robert thought about his answer for just a brief moment giving consideration to Charlie's comment. But then he grinned and chuckled, "You Yanks and your jokin', no wonder we let you all form your own country. We have genuine humor in Britain, Monte Python and of course The Royals. It don't get much more humorist than that. Fish with names. That's a good one."

Chapter 17
Tempest in a Teapot

Charlie spent the next few hours on the hotel's computer, researching what was happening with Elizabeth's case from the news standpoint. Much like other stories, they kept retelling the same main line. They even brought in so-called experts in the field of science and religion to be interviewed about the single point of the story. If a person was to hear the story and not know the truth, they would believe some terrible thing had been created. Truth was, they only had information that something has happened and the news people were making it the event of the day. A reporter knows how to twist and turn a simple line of a story into a full blown tempest in a teapot.

Charlie was reading a story online when he felt a hand on his shoulder. He looked up and it was Tucker. "You know, Charlie, they will infer a great many things into today's news story. Come tomorrow it will be mostly forgotten."

Charlie looked back at the screen, "Maybe, but somehow I don't think this is going away soon. I just can't figure how they got the story in the first place. I know that none of my people would leak this information. The fact is, they knew little of the details about her research. These reports tell of the possibility you have explained to me just today. They know the science, but not all of the reality in Elizabeth's case. I still don't understand a lot of the details. What I do know, it shouldn't affect her defense against the government's embezzlement charges. All these news reports will do is to cloud those issues at the heart of her case. People won't want to know about her guilt or innocence. They'll want to know about the news that she created life. This will most likely get out of hand quickly. I realize from the news reports, someone who knew how to put the information out there also knew how to keep it simple. From what I can tell they don't have much, just that she is credited with a discovery in genetics. The story's writer implied she created some sort of new life. A couple of the reports are even saying she helped to create a mammoth. After that, the story took on a life of its own. Every juror that will sit in judgment of this case will very soon hear about her case. Online news bloggers will twist and turn this case to create the most sensational news story possible. They will imply unrealistic facts into the case. Before this is done, some checkout tabloid will have a front page picture of a baby mammoth created by Elizabeth. My experience in a high profile case…when this happens, there is a good chance people will have a preconceived idea of her guilt or innocence. Some may even see this as an attack on their faith or their belief in mankind's science of discovery. Someone from inside Elizabeth's own research team had to have leaked this information."

Tucker sat down, "Take one step at a time. Don't go back home with preconceived ideas of what you'll find. Stay on mission to find the facts in your embezzlement case. The other story will be driven not by you or your team. If you give it any credence…well, it will be like stepping in dog poop. You'll track it everywhere you go and not be able to get the smell off your shoes. It will taint the case. Inside that courtroom the law is the guiding force. Best if you are not the one that tracks it into the courtroom."

Charlie looked at Tucker, "Is that your divine advice from above?"

"No, that's advice my mom gave me when I was a lad and stepped into dog poop and tracked it into her kitchen."

Both men laughed at his comment.

Charlie looked at his watch, "I'll get my things and check out. It will take a couple of hours at this time of day to get to the airport. By the time we get through security it will still give us a couple hours before our plane leaves. We can get something to eat. I don't think I can fly back to America on a couple bags of peanuts."

Tucker gave him a wink, "My friend, you won't have to. I took the liberty of upgrading your ticket to be in a seat beside me in first class. One of the benefits of being the head of a department at Stonemore is I get to fly first class. And in first class they serve a little better bag of peanuts."

Charlie was somewhat surprised at his generosity. "Well I thank you for that. That is downright generous of you."

Tucker stood and said, "We need to get going, and I told Robert to meet us out front. Damned hard to get a taxi at this time of day. Oops, I mean darn hard." The two men laughed as they prepared to leave.

Robert's driving gave little doubt he had maneuvered around the congested traffic of the airport many times before. As they pulled up to the unloading zone, he quickly exited and helped the two men with their single carry-on bags. Charlie extended his hand to Robert.

"If you are ever in my part of America, I'd like to introduce you to a fish named The General and then take you to a real football game."

Robert smiled as he said. "I look forward to a meeting with your famous fish. It will give me a lot to talk about when I return home. And as to your football… I'll reserve my judgment until I see a game."

A police officer stepped close and said to Robert, "Sir, you'll need to move along, you're holding up others who need to be unloading." Robert shook his head that he agreed as he gave Charlie one last handshake.

As he turned towards his car he suddenly remembered something. "Tucker, remember what Angus told you about the second set of samples." With that he waved and returned to his car.

It took almost two hours to make it through security and into the departure terminal. Charlie and Tucker found a coffee shop and made small talk for the next couple of hours. Finally the announcement came over the P.A. system for their flight to board.

It seemed to take no time once the two men were on board, until the plane was in the air. A young attendant made his way along the rows of seats asking, "May I get you gentlemen something to drink. Tucker responded, "A hot cup of tea. And do you have any biscuits?"

The young man sat a cup of tea on his table, followed by two biscuits.

"Sir, can I get you something?" he asked Charlie.

Charlie replied, "Coffee."

The man poured him a cup of coffee and moved on to the next row of passengers. Before long the two men were comfortable in their first class seats.

Charlie finally asked, "Tucker if you don't mind me asking, how did you get into the religion business?"

Tucker paused. "Well... when I was a lad I wanted to go into the army like my dad and his father before him. Kind of a family tradition. Growing up I heard stories of the first big world war from my grandfather. When grandfather's stories got old, my dad would tell about his war. I enjoyed hearing their stories, even though they were not all pleasant to hear. They really didn't talk about what they did, but more about the reality of war. They often talked about the loneliness and it was as big an enemy as the Germans were. I realized that their wars were not medals and glory. Still, in my heart I wanted to be a soldier. On the other hand my mum's faith was just as big an influence on my life. She took me to church and I felt a comfort in the presence of God. For a young lad growing up, I was torn between my desire to go into the army and a desire to be strong in my faith. My grandfather one Sunday recalled a story that helped me find the answer. He told of a vicar that came into the trenches in France. He trudged through the mud and horribleness of war to give mass to the soldiers. He told me he believed that vicar was the bravest man he had ever known. Even as the bombs were falling all around them, this vicar never flinched as he said mass and gave communion. After he finished he moved on to the next group of soldiers. My grandfather asked him. *"Aren't you afraid?"* He responded, *"No, I am in the army of God, He has sent me on a mission to spread His word in this place of despair and fear. If I am called home on this mission, He will have known I brought His words into the darkness of the hell on earth."* I realized that I could serve in the army but also I could serve in the army of God. So, the rest is history as they say. I spent twenty years in the army. I never had a mission in a place like that vicar in the First World War, but I still did my part to bring faith and comfort to soldiers far from home. One thing the army gave me was experience with other religions. I traveled the world and met people of other faiths, Islam, Hindus, Orthodox Christian, Buddhism, Taoism, Judaism and many more. When I got out of the service I moved from congregation to congregation for a while. Then I decided to go back to school at Stonemore and learn more about those religions I had experienced while in the service. After a couple years the university offered me a position. I took it and that was fifteen years ago. Now I am head of the department."

Charlie asked, "Ever been married? Or maybe your faith does not allow you to marry?"

"No, I never had the time to get married. And yes, my faith does allow its vicars to marry. I guess I just never took the time. As I get older I have thought about my choices more. I have no regrets. I remember those young men who were away from their families far from home. They were often scared and felt alone. I think I made a difference in bringing a little comfort to them. In a sense they were all my family. I take some comfort that I gave them a sense of family far from home. A couple of years ago a grown man with his wife and three children came into my office. I will have to confess, I did not recognize any of them. The grown man began by telling me he was one of the soldiers I comforted with faith in a faraway place. He always told himself someday he would find me and introduce his family. I guess seeing that man's family and hearing his words that I made a difference was as great as any reward I could have received in this world. I chose the way of faith many years ago and it has never let me down yet."

He touched Charlie's arm, "I know you don't have the faith that I have, but you need to know that I believe this will all work out for the best. I can tell you are deeply troubled by the recent events in your case. I have looked for answers ever since I first learned of Elizabeth's research. Angus helped her and he has spent many nights with the bottle trying to forget what they discovered. That reminds me, I am supposed to tell you that shortly after they received the first shipment of DNA samples, a second canister showed up. It also contained DNA, but no one seems to know anything about it. Angus wanted you to know."

Tucker paused and then finished his story, "As I told you earlier, I still believe God has a hand in all of this. We just may not understand how or why."

The two men finished their drinks and the host removed their cups. Charlie said to Tucker, "We better try and get a couple hours sleep, the time change will be hard enough even with some sleep."

The men turned off the overhead lights, the sounds of the planes engines soon gave way for them to sleep.

"Ladies and gentleman, we are beginning our approach to Washington Dulles. It is a pleasant seventy-two degrees upon our arrival. I ask at this time to place your seats in an upright position and prepare for landing."

Charlie was already awake looking out the window. His mind was working the case even before he returned home. In a couple of minutes Tucker asked, "How long will it take to fly from Dulles to, what's the name of the town?"

"Clarksburg. It shouldn't take more than an hour. Small regional airport. My friend Jake should meet us there, I texted him of our arrival time."

Before long the plane was lowering its landing wheels and the passengers could feel the plane slow as it increased its angle toward the runway. In a few minutes the lights of the city near the airport could be seen. From below the plane came the sudden screech of tires hitting the runway as the plane touched down. The plane shuttered as it slowed. In just a few moments it was taxiing on the runway. The overhead lights came on and people began getting ready to disembark. Charlie and Tucker sat for a few minutes letting those in a great rush move toward the front.

It was another fifteen minutes until the plane was at the gate and the door opened. Charlie and Tucker made small talk while the others moved toward the front. Finally the two men made their way to the door and onto the gate ramp.

As they came into the terminal building Charlie noticed three men in dark suits looking at them. One man pointed as they moved away from the crowded gate area.

"Mr. Charles Morgan?" one of the men asked.

"Yes, that's me. Can I help you gentlemen?"

"Mr. Morgan, can you please come with us?"

"And just who are you and these other men? If you know who I am, then you know I used to work for the Justice Department."

"Sir, we are with the State Department. We need to talk with you."

"I have a connecting flight I need to make. Is this going to take long?"

"Sir, can you please just come with us?"

Charlie resisted a bit, "No, you wait. If you have cause to detain me, just say it."

"Mr. Morgan, what we have to tell you may be beneficial to you, as it is to us."

Charlie turned to Tucker, "It seems I have little choice in this matter." Charlie looked at the man whom seemed to be in charge, "Can my friend accompany us?"

He looked at Tucker, "I see no reason he cannot if it is alright with you. Please follow me."

In a couple minutes they entered what looked to be an interrogation room. Charlie noticed the room had one mirrored wall. "Is this some type of interrogation?"

All three of the men in suits sat down at the table. "Mr. Morgan, please, would you and your friend sit. We have no intention of delaying you making your connecting flight."

Charlie and Tucker sat down and the tension in the room seamed to relax a little.

"My name is Conrad Jenkins, I am with the State Department and I suppose you know what is going on with your case back here in the states. Our only interest is, do you know or have any knowledge of the whereabouts of DNA samples that were in the possession of your client. We know she sent them to England last year after she obtained them. We have lost track of them since that time. You may or may not know the Israeli government wants the samples returned. We hope we can take care of this matter outside the scrutiny of the media. In fact, if the samples can be returned we can talk with the Justice

Department and see if the doctor's case may be handled in another way. This would save everyone a lot of problems. Also, there is the matter of a cryogenic storage chamber. It is possible this chamber may have some association with the samples."

Charlie sat back in his chair and then looked at Tucker before answering. "Mr. Jenkins, I hate to tell you this, but those samples you ask about are no longer available. In fact it is my understanding they have been scientifically processed. Why the big rush to gain access to those samples? It is my understanding these were legally obtained through proper channels. And they never entered this country before being processed. The State Department should have no interest in their whereabouts."

The man hesitated for a moment. He then leaned close to the agent on his right and whispered something. The man then stood and left the room. Charlie could see him calling someone on his phone.

"Mr. Morgan... Charlie, we have a situation. I do not know the circumstances under which Doctor Mathenson obtained the samples. All I know is the Israelis want them back. The highest levels of their government have protested their being removed by an American doctor. Now you tell me it is not possible to regain them. This creates a situation for your government and your client."

Charlie leaned across the table, "Let's be clear here, it may create a problem for you, but it does not for Doctor Mathenson. She broke no law. She's not been charged with a crime relating to the samples and I do not think the Israeli government or even our government want the world to know about these samples. Not that they are illegal, but, if you tell the world she had them, someone is going to ask why and what happened to them? Where did they go?"

Charlie stood up and walked to where the two way mirror was built into the wall. He pointed toward the mirror. "I'm bettin' beyond the glass is your boss. You tell him for my part and my client's part this is all. Let me repeat, ALL about the charges of embezzlement my client has been charged with. All the rest of what's going on is about something else and is your problem. The one thing I have come to understand in the last couple of days is this stinks of dog shit." He gave a quick look and winked at Tucker.

Charlie chuckled to himself before he began, "Recently I was told a simple story that fits this situation. Somebody has stepped in something that stinks and now everyone, except me and my client, seems to be trying to wipe it off their shoes. I don't have the sample, and as far as I know they don't exist any longer, at least in the form the Israeli's want back. Now, unless you have something else, my friend and I have a plane to catch."

With that the man doing the talking looked toward the mirrored glass. He stared for a moment. Before long the door opened. Charlie recognized the distinguished man from long ago. Back then Mitch Hamilton was a new assistant U.S. Attorney who worked for Charlie. Today he was The Whitehouse's Attorney General.

He offered his hand, "Charlie, what has it been, ten, fifteen years? If memory serves me, I assisted you in a case against a defense contractor. He had cost overruns of nearly a billion dollars in a five year period. While all those overruns were happening he purchased two houses in the Bahamas and one in Hawaii. No one could figure out where or how the money was missing from his project funds. But, you took a cocky full-of-himself CEO and picked him apart on the witness stand with his own words."

Charlie thought for a second before answering. "I remember. We got a conviction and the department cut him a deal. His political connections paid off. Too many senators getting major election contributions and whatever else goes along with having a rich defense contractor as a friend. Yea, I remember that. It was nearly twenty years ago. I also remember as soon as that case was over you called a press conference with reporters telling them how you cracked the case and saved taxpayer money. Just how much did you recover for the taxpayers?"

Hamilton said, "That's a long time ago. Let's get down to business. I'll have to figure out how to handle the missing samples with the Israeli government. The religious hard liners are using it in their country to make headway with trying to influence their government. They want to change some of their ways of dealing with those who are moving further away from the core beliefs of the Jewish faith. I'll guess I can figure it out somehow. Can you guarantee me that the samples are no longer available? I have to ask this, if the DNA was processed and no longer available, just what does that mean?"

Charlie looked at Hamilton, "Just what is the Attorney General of the United States doing handling a problem like this? And besides that, if there is something to be done it is a State Department matter. Someone in your position does not come out of his Washington office to talk with me in a back room of an airport. I can tell you want this to be totally unofficial that we even met."

Hamilton replied, "Charlie, you treated me with respect when I first worked with you, even though I was new to the job. My being here is my attempt to return that favor. All I want to know is the samples really gone and what happened to them? If they are, we'll figure some way to take care of this."

Tucker, who had sat quietly, said at this point, "Mr. Hamilton. I can assure you they are long ago processed at our university science department. We had no questions as to their origin or their proper ownership. Doctor Mathenson had all the proper documentations in hand before we began analyzing any of the samples. I was part of the process making sure all ethical questions were answered before the process was begun.

Stonemore is very careful to follow proper guidelines when it pertains to medical research, especially when that research pertains to the human race. As to where did the DNA samples go? They were analyzed using a special technique developed at Stonemore by your Doctor Mathenson and our Doctor Whitechurch. The material samples are rendered inert when they are processed. They are now data coded in a computer file."

Hamilton said, "Ok, let's say I believe you. What about the cryogenic chamber and its contents?"

Charlie looked at him and shrugged his shoulders, "I have no personal knowledge of this chamber or any of its contents. This question keeps coming up, but it is not something I have any knowledge of. Surely the government must have some idea about it. Why else would you have included it in the warrants that were issued? Where did you get your information?"

Hamilton said, "I cannot reveal that. But, let me say we are sure that it does exist and it is important to return it for several reasons."

"What is in that chamber?" Charlie asked pointedly.

Hamilton folded his arms as he walked around the room for just a moment. Then he said, "I don't know for sure, but a tip told us it has frozen embryos. Human eggs. Eggs that have been modified using the process Doctor Mathenson has supposedly perfected. Somehow the missing DNA figures into the embryos. And now we have this news report of her creating life. This is getting out of hand. Your government wants to put a stop to this now, quietly. Understand?"

Charlie shook his head as he walked around the small room. "Let me get this right. You'll forgive me, but I didn't take advance biogenetic science in school. I just got a crash course in the last few days. You're telling me you think Elizabeth Mathenson has somehow taken the information she obtained from those DNA specimens from Israel and somehow placed them in human eggs and now has viable embryos? That's impossible. They were reduced to computer data. No physical form to put inside your missing eggs."

Hamilton started to say, "We think........"

Tucker interrupted him, "Charlie, that may not be true. Elizabeth had an idea it may be possible, but she needed a lot more than she had at Stonemore to work with. She would need a source of human eggs along with a lab willing to remove the egg's nucleus and replace them with her synthetic DNA. When she left England she had no viable eggs. Are you saying she has come into possession of human eggs?"

Hamilton answered, "We think she purchased them a while back when a fly-by-night cryogenic storage lab closed down, went bankrupt. She placed her bid online with the auction house. She paid extra to have them shipped right away. She knew what she was doing. I mean she made sure the shipper understood the canister had to stay connected to the nitrogen supply. The foreclosure company itself did not realize what the lab canisters contained. They had an inventory list of abandoned medical equipment inside the storage building. They wanted to recover their financial investment by liquidating the equipment. The inventory only listed stainless steel storage containers. No reference to contents. The auction house moved quickly to sell off most of the contents in storage to recover their money. Items that went unsold were sent to a scrapyard and a landfill. No one gave any thought to what was in the storage canisters. They had no reason to. The items were advertised prior to auction as used medical equipment in storage. Records were not kept up to date, except to say two of the canisters were connected to nitrogen supplies and would be sold as a unit. That should have been a giveaway that they contained some biological material. But you have to remember, those doing the liquidation just believed it was old medical junk. No one seemed to pay attention at the time. It was not until the guy who bought the second chamber wanted to turn his into a backyard beer cooler. It was still connected to a nitrogen cylinder. He figured it would look neat in his backyard keeping his beer cold. Anyway, when he opened it he wasn't sure what was inside. A doctor who lived next door told him what he had purchased. He understood the sensitive nature of what the chamber contained and contacted the state health department. From there it escalated quickly. All of the items sold have been accounted for, except for the second cryogenic chamber and nitrogen cylinder. That particular chamber belonged to a fertilization clinic. The clinic had not been notified that the storage lab had been closed. Those two canisters were sent to the storage operation because the frozen embryos were no longer wanted by their owners. They were stored there until the clinic could properly deal with them. Somehow they were forgotten. The clinic was not notified the storage operation was being closed and its contents sold off in a bankruptcy deal, at least not until it was too late. The missing chamber contained one hundred and twenty five eggs. It's not known if they were kept sustained in their frozen state. If Mathenson has that chamber and the eggs were still viable, she may have transferred her synthetic DNA into those eggs. The religious world will go nuts. At the same time the science community would proclaim a great victory over the creation of man. There is no version where this comes out good for any of us. We need to find that chamber and end this madness. We still can do this while there is plausible deniability, at least officially. It will give us time to find answers. Because if she has done this thing, it won't be long before someone else is selling the technology on the internet. "

Tucker looked at Charlie, "If what he is saying is true, you cannot let them destroy those embryos. I don't know what the answer is, but destroying them is not the answer."

Tucker turned to Hamilton, "You don't have any idea where they are at this time?"

Hamilton answered, "No, the truth is, we hoped Elizabeth would work with us. There are irregularities in her spending on her research programs. And there are also questions we want answers to about her travels to Korea and Africa. Her indictment went forward in the first place because she failed to cooperate with answers to questions the government had about her program. You may think the indictment is about our pressing her to cooperate, I will tell you inside this room that was part of it. I'll deny I said that when we leave here. Then, we got a tip the hard records and computer files along with the cryogenic chamber were missing. A quick internal audit revealed they were in fact missing. Elizabeth was asked and wouldn't acknowledge she knew anything about their disappearance. And the worst part, there is someone inside her research team that's feeding information to an online news site for whistleblowers. Probably the same person who tipped my office in the first place. We can't keep this contained for much longer."

Charlie asked, "Maybe if you can find the whistleblower, they could give you more information."

"We don't know who they are, that's the point with these online news bloggers. Who can tell real news from fake news these days? Snippets of information sent to the site, and the person who sent it disappears. Our only hope is to convince Elizabeth to give us the chamber and the information of the entire process. We do know the chamber isn't something you can keep in your basement or garage. It has to be attached to a source of nitrogen and monitored. It is fairly large and bulky. She did not move it by herself. We have checked, she has no storage buildings or basements. No rental property. And no supplier of nitrogen has any records of delivering nitrogen to any place not accounted for in their records."

Charlie thought for a moment, "I can talk with her. If she cooperates this all goes away and she is reinstated with tenure and all rights? Is that what I can tell her?"

Hamilton, "Yes, but don't forget the other conditions Harrison told you at the pretrial hearing. There are questions that need answering."

Charlie looked at his watch, "Shit, we have missed our plane to Clarksburg."

Hamilton spoke quickly, "No, it is still at the gate. I think you'll find they had a mechanical problem. That problem will be taken care of as soon as you are on board. Being an attorney general does give me some clout."

Hamilton handed Charlie a business card, "That's my personal number, I will answer the phone any time you call. I can't stress to you the importance of this matter. I don't give a damn about the Israelis at this point. I do care about the questions this raises for science and religion. The government doesn't want to get into the middle of this question."

He looked at Tucker, "And I am not even sure, Father, your people want to get into this issue at this time."

Tucker smiled, "My people know the answer already. All life is precious. For us the how or why has already been answered. Still this raises questions that the church cannot stick its head into the sand and ignore. Simply putting our faith in God's hands is one thing, but when the church followers begin asking questions, we need to have answers they can understand. Blind faith beyond the clergy doesn't work these days. If we don't have the answers, they will quickly Google them."

Hamilton looked skeptical as he said, "Well, I hope your God can give us some help in controlling this situation before it gets out of hand."

Charlie and Tucker made their way to the small gate on the south end of the departure building. Two State Department men accompanied them. As they stepped up to the counter the one man showed them his credentials and told the ticket clerk, "These men are priority passengers."

She didn't seem to understand, but motioned Charlie and Tucker down the access hallway to the plane. It didn't take long and the plane was taxiing into position for takeoff. A moment later the captain announced, "Ladies and gentlemen welcome aboard. The temperature in Clarksburg is seventy-five degrees and our flight time is fifty-eight minutes. Sit back and relax." With that, the engines came to life and they started down the runway.

Tucker looked out the window after a few minutes into the flight. Your country is green and seems to have more roads than the whole of England. Charlie shook his head in agreement.

Chapter 18
New Ballgame

Charlie and Tucker's seats were near the rear. The sound of the engines made it difficult to hold a conversation. They did not want those around them to hear their conversation. Charlie hinted to Tucker any talk of the case would have to wait until they were on the ground.

It didn't seem like it had been nearly an hour when the captain once again came on the P.A. system. "Please fasten your seat belts and thank you for flying Green Mountain Airlines today." With that the plane began its descent into the airport.

It did not take long until they were next to the terminal building and exiting the plane. Charlie saw Jake standing near the fence.

Soon they made their way into the parking lot. "Jake, this is Tucker Hunter. He's a friend of Angus Whitechurch. He's returned with me to see firsthand what kind of a mess we have gotten ourselves into."

Jake shook Tucker's hand, "Father... or is it pastor?"

"Pastor will do just fine, but I'd like it if you just call me Tucker."

Fine…well, Tucker, you realize you may have gotten yourself into a world of, excuse the expression, shit!"

Tucker looked at Charlie and grinned, "It is kind of peculiar how often poop has come up in our conversations today."

Jake gave him an odd look as to his comment. "Charlie, I don't know where to begin, you know after you left, things went bad. At first, Elizabeth was a great deal of help in going through the financial records. But, into the evening she kept asking where you were. I put her off for a while. But when I went out to get something to eat, I think Downing must have told her you had gone to Stonemore. After that she seemed to be very irritated. She didn't seemed to want to continue any further that evening. So we finished up and agreed to return the next morning. That's when all hell broke loose. The street was packed with reporters and news trucks. We finally made our way into the office, Downing and me. But no Elizabeth. We tried to contact her, but no luck. Finally around noon she called and said she was unable to get into the office, what was she to do next. That's when we decided to move our operations to your office in Elkins. And when we got to your office it was all quiet, not sure how long that will last. When Elizabeth joined us, she was better. Still she seems at times to be lost in her thoughts. Don't know what happened, but certainly something did."

Charlie asked, "What did you find out about Clay Gordon?"

Jake looked through some papers on the front seat of the SUV. "Yea, I did, here is everything I could find. You told me he was just a law professor at the university and had been there a long time. That's true, but the story of his life runs deep and dark. There is a hidden mystery attached to his life. I couldn't find a way to see what it is. After he graduated law school he spent four years in the Israeli Army. Couldn't find out what his job was while he was in. After he got out, he just up and left his home and family with no explanation and came to America. A few months after he arrived, he secured a position as an assistant professor working in international law. He's considered an expert in patent law. After a couple of years he was offered a full professorship. His records shows he worked hard at remaining just a professor. He's had several opportunities to advance to higher positions and even take high level positions at other universities. He turned them all down. The only interest he showed beyond his classes, he returns to Israel six or seven times a year. Doesn't visit with family when he goes home. Not sure if that's his choice or theirs. They are inflexible in their beliefs. If they were Amish I'd say they were shunning him. He often speaks at the university and does research dealing with Israeli patent laws. Has few friends and stays to himself most of the time. One odd thing, he gives every impression that when it comes to his faith, he is a faithful Jew. My research shows his family is part of an extreme Ultra-Orthodox branch of the Jewish faith. It is known as Hasidism. They are very conservative in their beliefs. Go figure, it rarely happens that someone from so dedicated a branch of the faith chooses to adopt a more mainstream interpretation of the faith. Beyond that not much to tell about the professor. Oh, one more thing, about a year ago he made three trips back to his country in five weeks. Gave no explanation to the university, just said it was unavoidable. I can find no records of his travel or meetings while he was there. No credit card receipts, hotel reservations, nothing. It seems he went home and disappeared for a couple of weeks."

Charlie took the piece of paper from Jake's hand and looked it over for a few minutes. After a time he pointed toward the dates of last year's trips. "Are these the weeks he went missing?"

Jake shook his head indicating they were. "Something else, when I made a request of the customs in Israel about Gordon's coming and goings, I got a strange response to my question. Their records showed he was on a diplomatic mission. Any further inquiries would have to go through the Foreign Diplomatic Department, they told me. When I finally got in contact with them, they told me there was no such person and I must be mistaken about my inquiry. When I pointed out I was directed to their agency by customs, they told me I was mistaken. They were stiff and formal in their responses. Charlie, it's getting late; we need to get back to Elkins. Downing is probably wondering where we are and I am guessing Elizabeth wants to talk with you."

Charlie looked at him, "I can tell you, I want to talk with her. She's played me or us since this first began. She knows more about this than she lets on. She told us about the research at Stonemore and the possibilities it could create human life some time down the road. But, after talking with Whitechurch, I believe she has done it. Now, her research is missing from the lab, including the cryogenic chamber. If it weren't for Downing, I would just head back to the mountains and forget this whole mess. But, I am in too far to stop now. One thing for sure, she has to give me some answers."

As they headed east toward Elkins, Tucker leaned forward and touched Charlie's shoulder, "I've been thinking about what Jake said about this guy, Gordon. And it reminds me of something I learned during my time in the army. I was stationed in the Middle East for a while. I got to know a couple agents who were part of a security group known as Mossad. One evening we were having a drink and the one agent told me he was tired of the fighting in the Middle East. He had an uncle in government who could get him an assignment as a foreign agent. He explained they sent sleeper agents around the world. They would become part of communities for long periods of time. They were not what you could call a spy. They were there to learn culture and how society functioned over long periods. It was said they sought out places where different cultures came together. What better place than a university? They are known for having a very diverse group of people from around the world. I would suspect your university is such a place. That drunken agent figured it was a good assignment, as long as he was willing to walk away from his life in Israel. I suspect the next day he regretted getting drunk and revealing to me that bit of information. I know his superiors would deny his statements."

Charlie looked back over the seat, "So, you are saying Gordon could be some sort of Israeli spy? There are no government secrets in a university."

Jake perked up as he said, "That's not correct. Just think how many engineering, science and medical patents are secured by universities each year. Some universities and colleges do research for the government and private business. Given the fact that Gordon is an expert on patent law it would give him the inside track as to what's being developed. Who better to see inside patent applications then the university's expert patent lawyer?"

Charlie's face showed he was thinking. "Ok, but why break cover after all these years and risk exposing what and who you are. If Gordon is as you say an embedded technology spy, he could have gathered the information and simply handed it over the fence to a waiting agent. It looks to me like he is making every effort to not only find the information his government wants, he is after those embryos. Why all of a sudden expose a thirty year sleeper, if that's what he really is. And then why put yourself on a government team that's prosecuting Elizabeth? That just raises the odds someone will take notice."

Jake said, "Maybe he figured if the government flushed Elizabeth's secrets out during the embezzlement case, he'd be there to retrieve it. That may have been possible as long as her case remained simply about the money. Now that's all changed. Gordon is going to have to put himself out in the light of day to get ahold of the information and those embryos. I don't know why, but he risks losing a lifetime of anonymity for those embryos. And there is a second possibility, maybe our government is keeping him close, so they know what he is up to. Two birds with one stone. Elizabeth's research and an Israeli spy."

Tucker, who had sat listening as Jake and Charlie talked, waited until they paused, "Gentleman, think about it. It all comes down to the DNA samples. There is something about the samples that started this whole thing. I have an old friend who is a rabbi that I have known for a very long time. He is considered to be a very wise man in his faith. He is not officially in a position of leadership, but he has great respect among those seeking his advice. I could contact him and ask if he knows."

Jake looked at Charlie with a positive look. "It just might give us some insight."

It took about an hour until they were back in Elkins at Charlie's office.

Charlie pointed towards the side of the building. "Pull around back. No sense in advertising someone is in my office."

They sat in the SUV for a few minutes, wanting to see if anyone was watching. Convinced they were safe for the moment, they exited the vehicle and quickly made their way into the back door.

Downing's voice welcomed them as they came in, "Damn Charlie, I am glad to see you! Law school never prepared me for this part of the law. Did you find anything out?"

Charlie shook his head that he had. Then asked "Where's Elizabeth?"

Downing pointed toward his office's conference room. "Your secretary set up in there. She has been very helpful in getting things moving along."

Charlie said, "Give me some time to speak with Elizabeth, her case is a whole new ball game and I want to know who's on first and what's on second in this game."

He made his way to the conference room. As he did Tucker asked Jake questioningly, "Who's on first? What's on second? I don't understand."

Jake chuckled as he responded, "It's a reference to a comedy routine from long ago by two guys named Bud Abbott and Lou Costello. Charlie sometimes uses off beat references when he's looking for answers. I'm guessing Elizabeth better have who's on first and what's on second answers for Charlie."

Charlie slid the pocket doors open into the conference room. Elizabeth looked up at him as he entered. "You should have told me you were going to Stonemore. I could have saved you a trip. Besides, you were the one that kept going on about this being only about the money. What was it you said earlier on, 'Show me the money.' Well, there is no money in England."

Charlie said nothing as he walked to the coffee pot. He turned and held a cup toward Elizabeth. She shook her head, indicating she did not want a cup.

Still he said nothing as he sat down. Elizabeth was becoming aggravated at his not speaking to her. Finally she blurted out. "Well did you learn anything from Whitechurch? I am assuming that's who you went to see."

Charlie still did not answer as he took a slow drink of coffee. After he finished he looked up at Elizabeth, "You must have made this cup of coffee. I've tasted Downing's coffee and it's weak. This is good."

Elizabeth sat down across from Charlie at the table. "All right, what do you want to ask me?"

Charlie thought for a moment, "Where do I begin? Lady you have gotten us into a world of trouble. You have problems with the university, our government and the Israeli government. Not to mention Downing and Jake's involvement. I don't want to know about the money. I want to know what's going on. I know about the work at Stonemore. I also have learned about the cryogenic canister of eggs you bought. I also know none of these things in themselves started out to be illegal. But, whatever you have done has started a fire storm."

Elizabeth looked down at the table for a moment before looking back at Charlie. "Well, I am sure you have some understanding of the pulse diode analysis of the DNA. If you remember, I had already told you about that. So, it was no secret to you. The Israeli DNA samples were analyzed and turned into data. I am not sure if Whitechurch told you about the second set of DNA samples that arrived unexpectedly. I brought back two hundred samples to my lab. I injected the Israel DNA into bacterium. The bacterium regenerated the DNA. But, it had no real structure. It was a collection of chromosomes and genes. Now, here's where the magic happened, I took that combination of genetic material and placed them in the A.D.A.M. synthetic ladder matrix. Enzymes began trying to reassemble the DNA into a usable configuration. It didn't work in all of the procedures, but it did in some."

Charlie interrupted, "Is that where the embryos came into the picture?"

"Yes, we removed the original cell nucleus and insert what A.D.A.M. created."

Charlie leaned forward, "Where are the embryos? The governments of two countries are looking for them and you are the only person who knows where they are."

Elizabeth sat back in her chair. "Let's say I do know where they are, what difference does this make in my federal case? The indictment says I am charged with the question of funding. That's all! Correct?"

Charlie quickly answered, "Yes, but you remember the warrant included that cryogenic chamber and the samples. You may be charged with embezzlement of funds, but it is a short step to include the canister. Your only chance of getting through this and going back to the way it was is to cooperate with the government and tell them where the chamber is."

Elizabeth stood up as she said, "No, I will not give up those embryos, at least not to the government of either country."

She walked closer to Charlie and sat down in the chair next to him. "You have to understand, if I give those up, nothing changes. The research goes on in the backrooms of labs all around the world. What I have done is opened Pandora's Box.

But, I know from my research I am far from being the only one who has looked inside. Science, religious leaders and governments know this science is here. All they want to do is kick the question down the road for someone else to answer. Others and I who work in the genetics community have begun to realize we need to ask, 'Where are we going.' You must understand we are tampering with the future of mankind. I have used a simple bacterium and my molecule to create viable embryos. Inside that frozen chamber are one hundred and twenty-four viable embryos. Can you imagine if those doctors who performed horrible experiments on Jews had this technology? No, I will not reveal my data or even if I have those embryos. You asked to defend me on the embezzlement charges, I didn't come looking for you. As my attorney, I have client attorney confidentiality regarding anything I tell you. So understand all I want from you is to defend me on the legal charges."

Charlie leaned back in his chair, "Forgive me Doctor Mathenson, you are no doctor, in fact you should change your name to Joan of Arc. You want to throw yourself on the fires of ignorance so everyone can see your sacrifice. You better wake up. When they lock you up in federal prison, you will disappear. The story of your wild experimentation on human DNA will be twisted into a modern day Frankenstein story. In six months you will be old news, no make that six weeks. And if you think your sacrifice has made a difference your wrong. Those with the loudest voices will write your epitaph."

She looked at Charlie, "You mean, you and the others are not on my side to even try and defend me?"

Before Charlie could answer, a voice from the open door said, "I will stand with you Doctor Mathenson. Whatever you have done, it is by the hand of God. You may take credit for your discoveries, but I believe all that we are and what we do is by the hand of God. But Charlie is right, your voice will be silenced by the roar for justice."

Elizebeth looked up at Tucker and said, "Father Hunter, I did not know you were here. Why have you come?"

Your friend Charlie has invited me to go fishing in the mountains with him. Tells me the trout in his country are big. I explained to him that Irish trout are some of the finest in the world, but I have not heard of a trout in Britain with the name or rank of General.

Elizabeth looked at him and then Charlie with an odd look. "Tucker, I have gotten myself in a difficult situation. But nevertheless I intend to see it through. I will not reveal the information everyone wants. And I am going to trial to clear my name of embezzlement charges. Don't forget I may be the only name on the indictment, but there are at least a dozen people who have worked closely with me. If I give up, they may go after them. Some have been my family since I began my research. I will not let them suffer because of the direction of my work."

For a time no one spoke. By then Jake and Downing had joined them in the room. Finally Jake said, "Elizabeth, how do the Israeli's figure into this?"

Elizabeth thought for a moment before answering, "To be honest, I don't know for sure. I was at a conference that was held at the Weizmann Institute of Science. I met several scientists who were doing research on genetics. During our conversations one of the scientists told me about their DNA research program. He explained they had so many duplicate DNA samples, they were running out of room to store them. The duplicates were wasting away in some dusty medical cold storage unit. I offered to buy some of the samples for my research into DNA. He agreed with little hesitation and the samples were shipped off to Stonemore to be analyzed. No problems or even any questions about the whole deal. Now, I find out they are unhappy about the arrangements they made with me. What's even stranger is the second canister with DNA samples that I received. The Israelis won't acknowledge they were sent by them and the bar code numbers match the first shipment. Two identical groups of DNA samples and the Israeli government is going all out to get the first shipment returned and they don't even want to acknowledge the existence of the second group. What sense does that make."

Charlie ignored the second shipment as he said, "From what I understand, they are more than unhappy with you. This has created diplomatic problems behind the doors of governments. They want them back."

Elizabeth answered, "That is not going to happen. Those first DNA samples have been processed. And I have told you what I have done with the ones I shipped to my lab."

Charlie asked Elizabeth, "What do you know about Clay Gordon? Did you know he is some kind of a spy for the Israeli government?"

Tucker interrupted, "Charlie we don't know that for absolute certainty, in all likelihood he is, but we better be careful when calling someone a spy for a foreign government."

Charlie hesitated before saying to Elizabeth, "Tucker is right, so let me ask the question this way, Clay may be a foreign agent, is there anything you can tell us that might shed some light on his part in all this?"

Elizabeth responded, "No, not really, you have to understand that I have kept my nose into my research for so long, I have taken little notice of the world around me. Clay was just a professor at the university that I have known for... I am not even sure how long. He's just always been around. We talk when our paths cross. Other than that he's just there. I did ask him last year about patents and the university's rights to them. I knew the path that my research was on. I also needed to protect the research from being stolen or taken away. Clay gave me little hope that I could protect my research from being lost to others. That's when I began figuring a way to secure it myself."

Charlie asked, "He's never approached you asking for any further information of any kind?"

Elizabeth answered, "No, not that I can think of. I knew he was of the Jewish faith. I noticed the knotted tassels on his belt. I knew they had something to with is faith. But other than that he was just there. Very little social life, kept to himself."

Jake said, "Those are called Tzitzit. I believe they wear four of them. Has something to do with their faith."

Downing at that point said to the group, "We need to figure where we go from here. Jake and I have gone over the evidence in the case against Elizabeth. There are areas where they could make a case that she signed off on questionable things. Things that if you were to look at them singularly they may not be much, but when you look at them in wider concept, you could conclude a jury may determine she has acted irresponsibly in overseeing the project funds. Is that enough to send her to jail? I don't know, but it certainly could."

Charlie looked at Elizabeth, "I need to know if you want to go on from here with what you just heard. We can go into court and take your chances with a jury. And understand, you could go to jail for a long time. We could contact the Attorney General and tell him he has a deal. And you give him what he wants. If we go into court and use the defense you were acting in the best scientific interest of the country, we have to convince the jury of that. He has to allow us to enter evidence about your research. At the same time the prosecution will be entering evidence you mishandled funds. Anyway we go about it, I am not sure what this will mean with your career and your tenure."

Elizabeth thought for a moment and looked at the men sitting around the table. Finally she said, "I guess that decision was made last summer when I knew I would lose control of my research at some point in the future if I did nothing. As to the question of the Israeli samples, there is nothing I can do with that now. I want to go to court, plead my case and try and tell the jury why I did what I did. But, I will not tell where those embryos are. Not until I see the world is ready to face the questions of DNA science."

Charlie looked around the room, "Alright, you all heard her. Any of you want out at this point. I'll tell you it's going to get rough from here on out. This case will draw worldwide attention. And if I am hearing Elizabeth correctly, I am guessing that's what she wants. She's ready to face the consequences, each of you have to do the same thing."

Downing was first to speak. "When I first met the doctor she challenged me to rethink my ideals about science. I'd not given into the possibility science could create life. I don't profess to believe in God. Nor had I thought about where we came from, monkeys or an amoeba. I was a believer in the law. But in the last few weeks I have come to understand the law is what separates us from those things who can't think either way. I am not out to make a name for myself, but win or lose, I'll still be a lawyer the day following this case. The only questions I have, will I have done my best for my client. For me it's that plain and simple."

Charlie looked at Jake. "Don't look at me for answers, I am only here until the crappie start biting at my secret fishing hole. I can tell you this, I would not have missed being part of this."

Charlie sat back in his chair and smiled at the people in the room. "Well, it looks like we go to trial and hope the jury is open minded as you all are."

Tucker looked at Charlie and asked, "What do you think?"

"When I started this case, I was doing a person from my past a favor. And the truth is, I couldn't even call him a close friend. We were at one time, but things change, people change. I changed. I had forgotten why I became a lawyer in the first place. And when the person who once had been my friend asked me to defend the doctor, I guess I was unhappy he asked and I had no good reason to tell him, no. I was filled with anger and emptiness from the loss of my wife and son. I told myself if I was unpleasant enough around people they would leave me alone to feel sorry for myself. I even took my anger out on fish."

The others in the room did not understand his reference to the fish, but it was not important. They knew he was ready to go into court and defend his client.

"What's next Charlie?" Downing asked.

Charlie responded, "Tomorrow I will go see Clay Gordon. His part in this is not clear. The best way to find answers is head on. I'll go up to the university and just ask. Tucker, would you like to come along? Maybe this evening you can call your rabbi friend. He may give you something we can use when we talk with Clay tomorrow." Tucker shook his head he would.

The hour was late and they all have had a long day. They agreed to see what tomorrow would bring.

Chapter 19

The Answers Are In the Sands of Time

Downing drove back to his place in Monongahela City and Elizabeth stayed the night with friends in Weston, a town not far away. Jake and Tucker stayed at Charlie's place.

The three men took the evening off and visited a steak house on the outskirts of town. The waitress had to ask them a couple times to hold their laughter down. Three men in their late sixties telling stories from a lifetime of living. It had been a while since they had enjoyed spending time talking of the old days. Although Tucker had spent much of his early life in the military and the rest in the priesthood, he had a great many stories to tell. In fact, if one didn't know better you may have thought part of his life was spent as a truck driver.

Charlie asked, "In all those years in the military, you never had a girlfriend? If I remember right, you said your religious order could marry, if they chose to."

Tucker replied, "Girlfriend? Well there was one a long time ago. It was my second summer at the seminary. I was sent to South Africa to perform mission work for the summer. That was in the late sixties. I was young and my clerical collar wasn't so tight that I could not appreciate a pretty girl. There was one, her name was Abby. Her father was a sugar and tobacco broker. She had the prettiest auburn hair you ever saw. We went out a couple times. Finally she asked me, "When does your band play next?" I was not sure what she was meaning. So I asked, what band? She told me she thought I was in a touring British band. She thought I was wearing a Nehru jacket like Paul McCartney. When I told her it was not a Nehru jacket, but a priest coat and collar she nearly fainted. She hadn't realized she was dating a priest. She was a free spirit, but not the most intelligent person I ever met. Although, I will have to confess, I thought about learning to play the guitar after that summer. When my mission work ended, I returned to the seminary in England. I asked for divine forgiveness for my thoughts and actions that summer. Well, maybe not that one thing we did on the beach, it's good not to kiss and tell everything. Even the Almighty doesn't need to know details." The men laughed at his story. The waitress quieted them once again.

After two hours of stories, adventures and disturbing other customers, the group returned to Charlie's house. Tucker and Jake settled into a couple of wicker chairs and Charlie sat in the swing.

After a few minutes of relaxing, Charlie said to Tucker, "You want to try and call your friend? It's near nine in the morning in Israel."

Tucker indicated that he was ready to make the call. Charlie pointed toward the phone in his study. For nearly an hour Jake and Charlie sat on the porch drinking coffee and waiting for Tucker to return. Finally they heard the sound of the screen door opening.

Tucker had loosened his collar and sat back into the porch swing. "Well, I have the story. The lab curator who sold Elizabeth the samples, no longer works for the government's research center. He was dismissed as soon as they discovered what he had done. I am guessing he took the fall for what happened, but it was really the oversight committee's fault as much as it was his. There had been a debate as to what to do with the thousands of unused samples that had been collected in the early days of the Jewish history research project. The samples should have been destroyed in some honored ceremony, but everyone kept putting it off thinking some day they would get around to doing something about the samples. About that time Elizabeth came along, a conversation ensued about the samples. She explained to the manager of the lab about her research and the unique DNA samples could be critical to developing new ways to help cure genetic diseases. He liked the idea and allowed her to purchase them. Some documentation was sent along with the container to Stonemore. The information was mostly collection dates and regions of the country they were collected from. Individual identities were coded, no actual names were attached to the samples. A few months later someone put the ID codes together with names. Those particular samples were from some of the oldest families in Israel. In fact a couple families are said to trace back to Abraham. A claim that is impossible to authenticate, but because of the families known lineage, few would question their claim. When they found out the DNA samples had been sold to a person of non-Jewish faith and for research purposes, the preverbal dog shit hit the fan. Between the irregularities dealing with the samples in the Jewish faith and the embarrassment of having the samples sold right out of the country, it has become a full blown diplomatic fiasco. If it were to become widely known, it could cause major problems inside the ruling governing party. I also asked if he knew anything about Clay Gordon. He had no knowledge of who he is. But, he did say the intelligence service did send people to live in other countries to feed back information about culture and people's ideas about the Jewish faith. They take how other countries think about the Jewish faith very seriously. Given their faith's history, I can't say that I blame them. Whether Gordon is an agent or not, there is no way of finding out for sure, but he suspects he is."

Charlie looked at his watch, "I did not realize it was this late. If we are going to get up and head up to the university tomorrow, we better get to bed."

Tucker interrupted as he said, "There is something else. The rabbi said it is rumored that some in the government want the samples preserved in the English Genetic Ark Project. They knew Angus is head of the project. They felt it is insurance for the future if something would happen to our world. Most said they were crazy, but I think that group is who sent the second set of DNA samples to Elizabeth, hoping she would preserve them with her science. There is no way those behind sending the samples would ever admit to doing such a thing, but it fits with what we know."

Again Charlie said. "It is late. Too much to think about for now. We have a big day ahead of us. All agreed and retired for the night.

Morning came and Charlie was up and ready to go by eight. The smell of coffee soon brought the others downstairs.

Charlie asked Jake, "What are you going to do today? Is the rebuttal to the evidence the government has prepared ready?"

Jake stirred his coffee with a spoon as he answered. "Yea, it is pretty well ready to go. Waiting for a few more pieces of information from the university's control officer. Nothing that would hold us up. As for today, well, Nancy and I thought we would begin placing the information and case files in some order in legal boxes. I am sure you'll want to go over them with Downing to familiarize yourself with the evidence and additional files. How long you figuring to be in Monongahela City?"

Charlie finished his coffee and placed the cup in the sink, "Don't know for sure. Hoping Clay will talk with me. By now he probably knows I have an idea he is some sort of an agent. If he wants to go all cloak and dagger he might not even acknowledge he knows anything. Then again he might talk with me. I thought I knew the man for the last twenty years. Now I am not so sure."

The two men could hear Tucker coming down the steps, "That coffee smells good. I hope I didn't hold you up with your plans today. The time difference from England to here is a little tricky to get used to. It will even be worse going back. Let me get a cup of coffee and I'll be ready to go."

In a few minutes Charlie and Tucker were heading up the road toward a possible meeting with Clay Gordon. Tucker was fascinated by the towns and countryside along the way.

About an hour into their trip Charlie asked, "Forgive my ignorance in asking, but what is your title, Father, priest, minister, pastor or vicar? In the short period of time I have known you I have heard you called all of those titles. Just what is your official title?"

Tucker chuckled at Charlie's question, "Well, I guess officially I am, the Vicar Hunter. The truth is I am all those titles and even a few more. Depends on who's asking. When I was in the army I was expected to minister to all faiths. Very few times did anyone ask or even seem to care about what my faith was. They were more interested in talking over a scripture or something bothering them in the real world. It was not uncommon for a young soldier separated from his new wife and baby to be more interested in my telling him his wife still loved him. I even learned a little Hebrew. I guess that's why when I got out of the service I decided to continue my studies in world religions. In fact, I'll bet you didn't know I have a doctorate in theology. No one calls me doctor. The people I have met are looking for guidance. I don't think they are worried about who's reading the road map to heaven. So you see Charlie, I am a multidenominational man of faith. I guess my parents have to get some of the credit for my diversity. My parents were missionaries from England that worked in New Guinea at a mission. Some of the people working there were from New Zealand and some from Australia. My speech dialect is a little of all those nations. When I was ministering to those in the service the lads felt like they were talking with someone from home. They just didn't know which home." The two men laughed at his comments.

Charlie looked over at him and said, "Don't forget you are also a member of a rock and roll band in a Nehru jacket. Could you get me an autograph from McCartney?" The two men laughed at Charlie's comment.

It was near eleven o'clock when they arrived at the campus parking lot outside the law building. Charlie sat for a minute collecting his thoughts as to what he was going to say to Clay. After a few minutes he looked at Tucker, "I am not sure how this is going to go."

Tucker responded, "Do you think it will bother him that I came along?"

Charlie shook his head. "No, I don't think so. What he suspects we know so far, he's not going to be worried about you. In fact, I believe he already knows you are here. I want you along to help him understand what happened to the DNA samples at Stonemore. I am hoping that collar will help to convince him you are telling the truth. He knows I am a lawyer and the truth sometimes is not always in the answer, but in the question you ask."

The two men made their way inside to the security desk. Charlie remembered the man from the last time he was here. He seemed to be as disinterested this time as he was the last time. "Yes, can I help you gentlemen?"

Charlie spoke up and said, "Yes, do you remember me from the other day? I was here to speak with Professor Clay Gordon. We would like to speak with him. It is important."

The man looked at him with little emotions or sign of interest in Charlie's story. He picked up the phone and punched in a few numbers. Charlie could hear it ring on the other end. On the third ring he heard a voice answer. The security guard looked at Charlie and said to the voice on the other end, "It's that guy that was here the other day, says his name is Charlie Morgan."

Charlie could hear the voice on the other end, but could not make out the words. Finally the security guard hung up the phone and looked at Charlie. Professor Gordon says he will come down and talk with you in a few minutes. Said take a seat." The guard pointed toward a row of chairs next to the door.

As they sat down Tucker looked at Charlie, "Do you think he'll come down. Or is he on his way out the back door?"

Charlie gave him a shrug of the shoulder, "Not sure. He has me at a disadvantage in this. If I try and go after him, the security guard might have to move and use his pepper spray and Taser on me. I hate to think I made him get out of his chair when he seems to be enjoying it so much."

Ten minutes passed. Then to Charlie's surprise he saw Clay coming down the steps toward the guard's station. He stopped as he came close. "Charlie, who's your friend?"

"Vicar Tucker Hunter from Stonemore University. He teaches world religion."

Clay looked suspiciously at him, "You also have some interest into Angus Whitechurch and Elizabeth Mathenson when it comes to this matter of the tampering with life. Isn't that true, pastor?"

Tucker stood and looked Clay in the eyes. "I guess you could frame my part in this that way, but I think I am more of an advisor in matters of the church and science. Doesn't your faith allow Rabbis that latitude to give guidance in matters of God in the modern world?"

Clay paused and then said, "Yes, I believe you know one of those rabbis. You spoke with him last night, if I am not mistaken. Did you learn anything about me you didn't expect?"

Charlie made a sound clearing his throat, "Gentlemen, let's not have a pissing contest here in the lobby. Alright Clay, you know our cards and they are on the table. You also know we need to talk."

For a moment Clay and Tucker were locked in a staring contest. Finally Clay said, "Let's go for a walk away from the building, too many people already may know something is going on. My government is not the only player in this game. But, let me assure you we have only one goal, return of those DNA samples."

Charlie did not answer Clay's words as he turned and moved toward the door that led outside. Once they were no longer in the busy building they looked around for somewhere they could talk and not be disturbed. Charlie pointed toward an area where trees shaded several picnic style tables. Between classes students sometimes sat and worked.

"There, how about over there?" He said, Clay shook his head in agreement.

In a few minutes the three men made their way to the tree covered area and sat down at a table. Charlie began right away, "Why didn't you tell me of your involvement when we talked the other day? You let me think you knew very little about this case. Then I see you at the preliminary hearing and find out your some kind of a half ass spy. You know me, before I got involved in defending Elizabeth, I had no interest in any of this other shit that came along with this case."

Clay paused and sighed as he began, "Charlie, I will have to say I was not happy when I heard you were going to be part of this case. I figured Elizabeth would be represented by the green kid lawyer. The prosecutor would offer a deal and he would recommend it. I hoped she would take it. After all, her career and tenure were on the line. The gains versus the loss, I felt she would save her career and we'd all move on. But, then you and Jake showed up and couldn't let things go. No, you two had to dig for the truth. Hasn't working for the justice department all those years taught you justice is in the eye of the beholder of the real truth. Elizabeth has not only violated the trust of the legal system, but also God. It wasn't bad enough she was running around the world looking for the secret to creation of dead mammoths. No, she had to go into my country and take the very essence of my people. For thousands of years outsiders have sought to destroy my people. Now, in the name of science, she follows in the footsteps of all those who have tried to oppress my faith. Her transgressions against the Jewish faith cannot go unchallenged or unpunished." Gordon was visibly angry.

Tucker interrupted him. "How can you say she has done all that? You as a man of the law and knowledge knows she didn't plan any of this? It's not like she broke in the back door and stole those samples. Her intentions are purely science. Your suspicious minds have created this situation."

Clay leaned across the table and looked at Charlie, "Where are the embryos she created with our people's DNA?"

Charlie snapped his response, "I have no idea of their whereabouts or even if they really do exist. People keep asking about them, but I have yet to see any evidence they really do exist. Besides, you need to remember I am representing her in a federal embezzlement case. I am guessing you are the inside whistle blower that started all this in the first place."

Clay laughed at Charlie's comments. "You fool. She has played you since the beginning. I don't think she knew you were going to come along, but when you did she had to know it would raise the profile of this case in the media. Besides, if you hadn't gotten involved, she would most likely have taken the prosecutor's offer. I don't think she was willing to give up her career for the right to keep the secrets to her discovery. That green lawyer would have heard the offer from the prosecution. If I am right, he would have weighed the deal versus the possible outcome of a trial. He would recommend she take it. Jail time verses lifelong career? I think she's a smart woman.

Everybody gets what they want. My government would have gotten what they wanted. The university would have also gotten what they wanted. And I could have gone back to being a simple law professor. Now that is unlikely to happen. Me a whistle blower, you got it all wrong. If I had done that, it would have exposed my interest in the case. My position with my government is no longer a secret. In fact, you, an old retired lawyer, even knows who and what I am."

Charlie sat for a moment thinking of what Clay had said. "Is that why you sent me that note in Hebrew as a warning?"

Clay for the first time got a strange look on his face. "What the hell are you talking about? I never sent you or anyone else a letter written in Hebrew or any other language. You're imagining things Charlie. There is more at stake here than you know. If I were you, I would be careful. That cryogenic chamber is worth a lot to some people who want the technology. I only want the chamber to protect my people. There are others who want it for reasons they only talk about behind boardroom close doors."

Before Clay could say anything else he pointed towards a young woman coming their way, "Well, Charlie, here comes a reporter who wants answers. Maybe she can tell you where you are going from here. Maybe she'll even know who sent you that letter. You'll excuse me, I have a class to teach."

As he stood up he leaned down toward Charlie, "Tell Elizabeth to take the damned prosecutions deal and put an end to this before it gets out of hand. It is time for her to stop playing God."

In a moment a woman in high heels approached the men's location. She was having trouble walking on the soft grass of the lawn. As she came closer, she was already talking into the microphone that she held in her hand. She was trailed close by a man with a video camera on his shoulder. "Sir aren't you Charlie

Morgan, one of the lawyers representing Doctor Elizabeth Mathenson?"

Charlie looked over at Tucker, "I think it is time for us to go."

As he stood to leave, the attractive young news reporter asked, "Is it true your client has found a way to create life? And if that is true, how does she feel about the Vatican this morning saying this discovery is some grand conspiracy to lessen the importance of the church? I also understand the doctor has chosen to use people whose families go back into biblical history? Is she using her controversial experiments to offset any critics from the Vatican or Jerusalem?"

Charlie looked at Tucker as they walked away, "Did I ever tell you my life as a fisherman was boring? I liked boring and being left alone. I did not have to answer questions like this when I came out of a mountain stream."

Tucker replied, "That may be true Charlie, but somehow I think you are going to have to answer these questions for yourself first, then in the eyes of the world. I can't say I envy you that task. But, if I can be of any help, I'll try. All you have to do is ask. These people are not going to take an answer from the Bible. They smell a ratings story of historical proportions. The words in the Bible will be something they use to sensationalize the story. Their side will be Genesis. Your side won't be Darwin's theory of evolution, but Elizabeth's theory of creating life by the hands of mankind. Not even Charles Darwin could have predicted this. He believed time and nature would create life in all forms. Suddenly Doctor Elizabeth Mathenson and her science can decide what life gets created and what life dies out. And you, Charlie Morgan will defend her rights to replace God as the creator. After this is over, you will be the most famous lawyer in America. No, not just America, the entire world. Or should I say you may be the most infamous. You already have the Jews working against you. And on TV they say the Vatican is raising issues of faith about Elizabeth's science. And Downing tells me about some evangelistic preacher who held a rally with his congregation telling them to picket James office. Somehow your client has shaken the world churches and their faith."

Charlie looked at Tucker as they got into the SUV. "What about you? Is it fair to say you represent the Church of England? Do you believe it is as they say?"

"No, Charlie, I told you when I came along. God gave me intelligence to look at the world with eyes to question and seek understanding. If what they say about Elizabeth is true, I still have to believe it is only possible because this was meant to be. This is not a plot to replace God. She went looking for ways to better life in this world, to eliminate the suffering of so many who endure illness and physical affliction. But the road she went down had many destinations. She chose this one, not for evil intent, but seeking answers. That answer is a far distance from where she began her journey. Some will say her work is that of the Devil. It is simply the work of a scientist."

Tucker laid his hand on Charlie's shoulder, "I know some have decided it is a plot to recreate man in his own image. Before Elizabeth's science, the Israelites believed they were the chosen people and have a covenant with God. Now, will those born from this technology also have a covenant? Not with the God of Abraham, but with man. You see the very question of her science raises so many issues with their basic beliefs, those in power will never accept it. The truth of her science will never have a chance and neither will Elizabeth. There is a passage I tell my congregation when they question such things as you have in this case. In Hebrews 3:4 it is written, 'For every house is built by someone, but the builder of all things is God.' Charlie, I hesitate to say this, but I feel Elizabeth could be in danger."

Charlie drove several miles after Tucker's words. Finally he said, "You make a convincing argument. You should have been an attorney. Unfortunately a jury needs to hear more than, trust me, God has the answers. He just doesn't always tell us why. They will want real answers. Both you and I know that in order to prove this case, I will have to prove to the jury that she has not done this for greed or intent to create a woolly mammoth. The prosecutor in this case is Harrison Dobbs, he will try not to allow any of this into the courtroom. He wants his case to be straightforward, the letter of the law and she clearly broke it, period! He has high aspirations for his career. A high profile case deciding between science and God is too uncertain to benefit his career. Win with a decision based on clear facts she embezzled money and no one will pay any attention. But, win a case based on the fact she broke God's law with that money and he places himself squarely in the conservative's camp. In a confirmation hearing he will have to say in front of cameras, he used his strong faith to find justice. A judge nominated to the highest bench in the land needs to have decisions which covers a known judicial career. Those decisions need to be an indication of their thinking. One decision could define his whole career. That may be good or bad. He can't take that risk with this case."

The two men rode further down the highway in silence. Finally Charlie said to Tucker. "In the console there should be a business card for Mitch Hamilton, he said call if I need anything. I think I should give him a call."

Tucker searched for a few moments through the console, finally he found it. "Is this it, the U.S. Attorney General's number?"

"Yea, that's it. Charlie pressed the phone button on the steering wheel, a voice asked for the number he wanted called. Charlie relayed the number to the unseen voice. After a few seconds the phone rang on the other end.

"Hello, this is Mitch Hamilton, who is this?"

"Mitch, this is Charlie Morgan. You said call if I needed anything. As you probably know this case is getting to a point where outside events are driving the story. Something neither of us wants. The Vatican and the Israeli government are protesting Elizabeth's case. She's received some anonymous threats. You and I have seen this before in a case, but his time it is a little different. Some of these people are on a mission from God. The last thing any of us want is for something to happen to the doctor. I need you to supply Elizabeth with some protection. At least until we get this settled."

Mitch hesitated for a moment, "Charlie, give me a moment; I am in a meeting. I'll put you on hold."

With that the phone went quiet as Charlie listened to the silence as he drove. Tucker said, "Do you think he knows what's going on?"

Charlie looked over at him, "I have no doubt. I'm betting the Justice Department and the State Department are trying to figure how to handle the Israelis and the Vatican."

About then Mitch came back on the car phone, "Charlie, are you still there?"

Charlie responded, "Yes, I am here. Do you think you can help Elizabeth?"

"Charlie, I have to tell you the best way to handle this for all of us is for her to give up the information and the location of the cryogenic chamber. That's the only way I can begin to make things better."

"Mitch, I don't think that is going to happen. Elizabeth is set on going to court. She's not going to reveal what you're asking. It's her life's work and she knows it's over as soon as she gives it to you or the Israelis. This case is going to go forward, no matter what. At this point I don't even think the Justice Department could drop charges and not have major repercussions in the media. The entire world is wanting to know, has this woman created a way to play God."

"What do you want from me Charlie?"

"A couple of Marshals, a while back, did some escort work that I was associated with. Both of them knew their jobs. I'd ask you to assign them to keep an eye on the doctor until this is settled."

On the other end of the phone was a long pause, then. "Alright Charlie, but you have put me in a difficult position. The Justice Department is prosecuting the woman and now we are providing protection for her. I'll have to answer for this."

Charlie spoke up quickly, "Well Mitch, what would you rather do, tell your Whitehouse boss that you ordered the Marshal service to protect this woman or tell him she had been injured or even worse? I don't think we have a real choice here. Too many players and the game's gotten more dangerous."

Charlie could hear Mitch rustling papers, "All right Charlie, give me those Marshals' names. And what about you and the others, are you also in danger?"

"U.S. Marshals Jud Richmond and Samantha Harris. They are out of the northern district office. As to me and Downing, we'll be okay. People guarding us would draw a lot of attention. No, for now, just watch over the doctor."

Mitch sighed over the phone, "All right Charlie, I'll take care of it. Now, I will ask you to please talk with your client and try and get her to work with us. We can protect her research and make this go away. In six months I can't say it will be back the way it was, but she will be back to doing research on her DNA medical science to help the sick, the goal she set for herself long ago. I'll talk with you later." The phone connection went dead.

Charlie looked at Tucker, "Well, at least we won't have to worry about Elizabeth's safety."

Tucker said in response to his comments, "That's true, but you and Downing are just as much at risk as she is. Why refuse the protection for yourselves?"

Charlie chuckled, "Well, I guess risk goes with the job. I have done this for a long time, never figured I was in danger before and I am too old to start worrying about it now. As for Downing, I'll talk with him and see how he feels about the danger."

Tucker then asked, "Have you decided to ask Elizabeth if she is playing you for a fool?"

Charlie thought of the question before he answered, "I've thought about it a good bit. I've argued it both ways in my mind. I'll not say anything at this time. If it is true, I'll know it, and saying something will not help the case. If it's not true, then if I talk with her it will just add to reasons for her mistrusting me. I have no doubt she knows about the missing records and the cryogenic chamber. For now it is enough we understand she has them. She has chosen not to acknowledge that fact, even when she knows we believe she has them. No… I'll play along, at least for now. I have taken on cases before when I was not getting all the truth and going into court. The trick is to know your client is holding information back on you. And remember my client was often the federal government." Both men got a laugh out of that.

Chapter 20
A Time to Judgement

The next two weeks passed fairly quickly. Charlie and Downing went over every aspect of their defense. They struggled with the list of potential witnesses they intended to call to rebut the prosecution's evidence. Even more important were witnesses needed to support their defense. Charlie knew the prosecution would have a fit with their late date of notification of intended witnesses. Dobbs' list of witnesses all were there because of their knowledge of the embezzlement charges against Elizabeth. He had not included any witnesses that could speak to the questions behind the case of how the money was spent. Charlie knew he was not going to give any cause to allow that information to be heard by the jury. Charlie also understood Dobbs was well versed in courtroom surprises. He was sure that he would have done his homework into her research with DNA and the problems it had caused. If he intended to try and keep it out, he would have to know as much about it as possible, to object to any motion the defense would raise. Charlie and Downing knew their first goal in her defense was to convince Judge Thompson of the importance to look behind the outer layers of the case. And even if he did, they would have to hope the jury would be understanding, beyond the prosecutor's evidence. The defense of her case relied on the jury being able to comprehend more than just the financial aspects of her research. They would have to prove to the jury her research evolved with each new discovery. And those discoveries were still within the scope of the original goals of the grant monies. Her intent was purely for advances in medicine, not uncontrolled science. At least not uncontrolled by Elizabeth.

While they were preparing their case for the court of law, the court of public opinion was in full session. Daily talk shows were debating the possibility that science had overreached its limits and was now treading on the morals of mankind. Newspapers ran stories of science gone amuck and lack of governmental regulations that could have prevented such an occurrence.

Television ministers were proclaiming to those who watched their programs, perhaps this is God's way of punishing us for the lack of true believers. Church attendance rose during the weeks after it was announced that science was on the verge of creating a man or woman.

The world had begun to judge the case even before the trial was to begin. Charlie realized it would be impossible to prevent this from happening.

Finally the day came to return to Wheeling's Federal Courthouse. As they rode in the SUV, Charlie reviewed with Elizabeth the procedures for a federal trial. He wanted to prepare her for how he expected the case would progress. He explained to her he did not want her to over react if she was surprised at anything that may be introduced. Their SUV was being driven by the U.S. Marshal Jud Hickman. The Attorney General had been good to his word and assigned them to Elizebeth. Jake, Tucker and U.S. Marshal Samantha Harris followed in a second SUV.

Charlie flipped through papers, marking sections he wanted to remind Downing to cover in his opening remarks. He looked over his glasses at Downing. "I anticipate with the amount of publicity this case has drawn in the last couple of weeks, the prosecution is likely to have revised how they are going to approach making their case. I still don't think they will bring up the issues behind the embezzlement case in their opening remarks. They are likely to stay on course with their original case, while all the time preparing to shift the intent of the case to the work you were doing at the university. Dobbs knows we will use your work to explain the use of the money. I don't think he will concentrate much on the work you and your teams have done in the university's labs. I think he will focus more on what you were doing in Korea, England and maybe even

Scotland. He will want to imply a connection with other governments and rogue science. He realized the only way to convince the jury of your intent will be to wait for you to take the stand in your own defense. I don't like having to place you on the witness stand, but I don't see that we have any choice. When it comes right down to it, those people on the jury will want to hear what you have to say. Their ideas and attitudes toward you will have been shaped by what they have read and seen on television. Most of that has been speculation. I'll bet there have been more news stories on DNA research in the last two weeks than has been in the last ten years. When you step into the courtroom everyone there will know you. Not by who you are, but by how the news has portrayed you. And I know you don't want to hear this again, but when Dobbs gets you on the stand, he's is going to ask you point blank, 'Where are the records and the cryogenic chamber?' You can take the fifth or tell them… well, whatever you decide. But that will be the moment of truth. Your testimony will be key to our case."

Elizabeth did not respond to the implied question of the missing items. "I am ready. But what about the other witnesses you intend to call?"

"Having met Angus Whitechurch, I believe he can bring an understanding to your work and how it has evolved as the science has progressed. With his testimony, we should be able to give the jury an understanding that you didn't decide to begin a new course of research, but the same one you started years ago. You need to keep in mind that the members of the jury are people from everyday walks of life. They will not be accountants nor scientists. But yet they are going to be asked to make a decision in your case by judging the information in two areas in which they have little understanding. What they have gotten they got off network news. I am hoping Angus's down-to-earth style and your expertise can convince them you spent money in pursuit of better medicine. It still won't answer the question of what they say you stole from your own project."

At that moment Marshal Hickman's phone rang. He pulled off the street and into a gas station lot. "Yes, this is Marshall Hickman."

For the next few minutes his conversation was about arrangements that had been made at the courthouse. Preparations were made to give access to the federal building from the backside. They were to enter into the lower secure parking lot. In front of the courthouse, cable news satellite trucks were nearly blocking the street. City police were trying to keep the traffic moving while making provisions for the trucks. Servants for God were carrying signs protesting Elizabeth's work. Several scuffles broke out between those who believed her work was that of the devil and those who believed her work deserved a Nobel Prize.

As they pulled onto the street they could see the crowds in front of the courthouse. Hickman spoke into his wrist communication device, advising his ground team he was on approach and would be turning into the alley one block south of the courthouse. Everything went smoothly. In a few minutes they were secure in the sub-basement.

Hickman turned to the three in the SUV. "All right this is how it is going to go down. I will open the door and assess the area. Two other marshals will quickly escort you to the elevator. It will take you to the floor where the courtroom is. Do not hesitate, do as the marshals tell you. Do you have any questions before we begin?"

They shook their heads they understood. Jud said "Let's go." He stepped out of the vehicle, looked around then opened the door for Elizabeth. "Move quickly to the elevator."

In a flash, Elizabeth, Downing and Charlie were on an elevator with three marshals heading toward the second floor. In the background, Charlie could hear playing very low, Barry Manilow. He mumbled to himself something unpleasant about the damn background noise.

The elevator stopped and the doors opened. They could hear a loud collection of voices in the hallway. The marshals stepped out and were greeted by two more. Hickman turned to the three in the elevator, "All right, you ready for this?"

Downing said without hesitation to the marshal, "Let's go."

Surrounded by marshals, the three were escorted into the crowded hallway where uniformed police cleared a path. Reporters were calling out questions as they moved toward the door.

Charlie holding Elizabeth's arm, spoke to her in a quiet tone, "Don't worry, we will be in the courtroom in just a moment. Don't pay any attention to reporters' questions. This is only the first day. You'll have plenty of time to speak when the time comes."

The marshals were very effective in moving the crowd away from the small group as they moved toward the door. In just a few moments they were in the courtroom. It was quiet. The prosecution's team was there along with their support staff. Four uniformed officers stood around the room. They all were stoned face and consistently monitoring the room, even with only the few that were allowed in so far.

Dobbs motioned to Charlie and Downing. Charlie nudged Downing, "He's wanting a side bar. Let's see what he wants."

Dobbs and one of his associates, Lisa Williams headed toward a room off the left side of the courtroom. It was a small room stacked with file boxes. "Alright Harrison, what do you want?"

Harrison leaned back against a filing cabinet. "Charlie, since we met a month ago, this case has taken a great may turns. I don't think you have any idea how much pressure has come to bear on aspects of the justice department to end this. In fact, it was suggested to postpone this case for a year until things quieted down. But, I don't think that is going to happen. Doctor Mathenson's case has raised a great many questions. I don't want answers to come outside of this courtroom. Here is what I suggest. She turns over all the information about her research to the university. Then she reveals the location of the cryogenic chamber to the State Department. If she does that it gets Israel and the Vatican off our backs. I want your client to get a fair trial. Do you think she can get a fair hearing with the environment that is swirling around this courthouse?"

Charlie looked at Downing before answering. "Harrison, I will speak with Doctor Mathenson once again about your offer. I do not believe she will accept it any more than the last time you made it. If, and I do mean if, she knows of the items you speak of, I don't thinks she's going to change her mind. She is willing to stand trial on the embezzlement charges. Elizabeth knows she has not been charged with theft or crimes against humanity as some have suggested. You think you have her between a rock and a hard place. Truth is, we all are in a tight place if this gets out of hand in the media."

Dobbs turned to his associate. The young woman handed him a silver DVD disk. "Charlie, as part of the disclosure of evidence, this was just made available to us. I think you should see what is on it. Show it to Elizabeth. Today is jury selection. I understand they have seventy-five people waiting to go through the process. What's she will see on the DVD may make a difference in the question of the cryogenic chamber issue."

Charlie took the DVD and without looking down at it handed it to Downing. "You say you just came by this? What does that mean?"

"Just what I said. The university went through all their security video and came up with this evidence. It took awhile because they didn't know for sure or how the cryogenic chamber was stolen. They knew it did not go through building security. It would have also required proper documentation to take it off site."

Charlie thought for a moment before he spoke, "Have you seen it?"

"Of course. It's clear what happened to the cryogenic chamber. Maybe it will spark Doctor Mathenson's memory after seeing this."

Charlie thought for a second. "Give me the short of it. What's on the disk?"

"It is footage from security cameras in the utility tunnel between the old section of the school hospital and Mathenson's research lab. Three months ago the camera captured two individuals moving what appears to be the canister along the tunnel toward the old hospital. They apparently used the tunnel to move it to the old basement. Once there they loaded it into a vehicle in the back parking lot. We found the hand truck they used to move it along the tunnel sitting next to the dock."

Downing spoke out, "So you have seen who took it out of the lab?"

Dobbs' face twisted a little before he answered. Charlie knew there was something about Downing's question Dobbs didn't want to tell them.

Charlie wasn't going to let it pass. "Well Harrison, answer his question, do you know who stole the cryogenic chamber?"

"Not exactly, when you watch, you will see the people in the video are wearing dark hoodies with heavily insulated jackets. We couldn't see their faces, but who else could it be? And one more thing. There were two cameras, one pointing into the back parking lot and one into the abandoned boiler room. Both cameras were not working. The fact is, a few weeks before they both stopped working. They are in low priory locations and campus security had not gotten around to replacing them. The odd thing is about how they were disabled. They showed that no one had physically touched them. But, when they opened them they saw the heat sink on the mother boards had been shattered like glass. The only way this could have happened, the technician believed they were exposed to liquid nitrogen. He believes someone poured the nitrogen into a small service port on the side. The liquid would have run in on the hot heat sink. Instant fracturing. The nitrogen would evaporate instantly leaving no trace. They have no explanation how this could have happened under normal circumstances. Someone had to plan to disable the cameras to prevent anyone seeing them loading the cryogenic chamber into a vehicle and driving away. More important, it would take someone who had been around a long time to know about the old service tunnel, a forgotten access that led out of the building without guards."

Downing started to say something. Charlie held his hand up to stop him. "Harrison, what you have here is meaningless, and you know it. It's a scare tactic to try and force our client into an admission of guilt. From what you tell me, this could have been you in the video and you couldn't tell. As to the destruction of the cameras, sounds to me like they have some faulty equipment. No mystery how they failed. That is unless in a court of law you want to imply our client is somehow guilty of this too. Video of two shadowy people in the night that may or may not show an alleged theft is hardly evidence against our client. You are grasping at straws here. Enter the video if you like. We find no objection to it being entered. It will just show you are reaching to get a conviction. Thompson will unlikely give you that much latitude, I am guessing. No, I'll let the jury decide if you choose to enter all this cloak and dagger stuff into evidence.

I think we are done here. Anything else about the deal you want me to tell our client?"

Dobbs stepped closer to Charlie. "Tell her she will lose in both the court of public opinion and in the courtroom. If she goes through with this she will have ten to twenty years in prison to think about it." He took his finger and pointed it toward Charlie's chest. "You tell her!!!"

Before he could pull it back Charlie grabbed it and held it. "Don't you dare try to intimidate me. I remember you when you could not even prepare simple brief for a hearing. And now you think you are some big shot prosecutor? Your ass is in a sling here and you know it."

Charlie leaned even closer to Dobbs, "Anyone talking about a Supreme Court appointment lately? I don't think so. You're pissed that you took on this case. Your views on the constitution may not be on public display, but your narrow minded ideas of the world will be. I have nothing to lose, I am an old lawyer simply protecting my client. You're a professional politician disguised as a lawyer." With that Charlie released his finger. Dobbs pulled it back quickly and the men stood toe to toe not speaking for twenty seconds. Finally the young attorney with Dobbs said, "Sir, we need to get back into the courtroom, they will be bringing the potential jurors in shortly."

Dobbs stepped back and adjusted his coat. Before turning for the door he glanced back at Charlie, "Tell your client of the deal, time is running out for all of us."

In a few minutes Charlie and Downing were back speaking with Elizabeth. "The government wants to make the offer one more time. But this time they may want to put off your trial for as much as a year. That would let things in the media quiet down. And they want you to turn over the records of your work to the university and the cryogenic chamber to the government. The same thing the government wanted in the first place. What do you want to do? Oh, I almost forgot." He showed her the security disk. "Dobbs says this shows the theft of the canister from your lab through some old tunnel that runs to the old hospital basement."

Charlie watched for a reaction in Elizabeth's eyes. She never flinched. "So, who does it show stealing the canister?" she asked.

"According to Dobbs, it's hard to tell. They wore hoodies and bulky coats. Can't really tell. But they know it was taken from the lab by the tunnel under the old hospital and loaded into a vehicle and taken away. He believes you are one of the people and he will try and sway the jury toward that conclusion. He never said anything about who he thought the other person was."

Elizabeth sat back in her chair unflinching in her expression. "Charlie, James, I do not want to make a deal, I am not guilty. It now seems the entire world is watching. If I cut a deal, they will believe I did it because I am guilty of embezzlement and didn't want to go to jail. I want the world to know I am not guilty. But, I want the world to know…"

Charlie said quickly, "To know what? If you expect me to defend you, you have to be honest with us. Our asses are on the line too. Especially James', he's at the beginning of his career."

Before she could answer, a court bailiff tapped Charlie on the shoulder. "Judge Thompson would like to see both counsels in his chamber before we begin."

Charlie looked at Dobbs who was getting the same message. In a couple of minutes all the attorneys were being escorted into the judge's chambers. The judge was standing looking out the window.

Without turning to look at the four attorneys, he said, "My granddaughter asked me a couple of days ago was I going to be the judge of the mammoth lady. My priest asked if I was the judge in the case of the doctor who claims to have created a man."

He turned to the group standing looking at him. "I told them both, it was complicated and the lady was not on trial for what they were saying on television."

He came around the desk and sighed deeply. "Charlie, what are we doing in this kind of case? Hell we both are retired and gave this all up."

Charlie looked at the judge. "You know better than that, we took an oath a long time ago to uphold the law. We knew back then sometimes that may not always be easy. And sometimes it would require that we put aside our own feelings to serve the law. After all, if we start picking and choosing which laws we are comfortable with, it lessens the whole system."

Judge Thompson said nothing as he returned to behind his desk. "All right Charlie and Harrison, you have been in my courtroom before. You know I don't like long winded presentations that go over and over and over the facts you have to present. Say what you have and move on. Due to the nature of this case and what it has become, I am going to place gag orders on everyone associated with this case. I intend to keep the case in the courtroom as much as possible. There is a large group of reporters out there, ready to dissect every word said. I can't prevent that. But those who are part of my court will not add to the news reports. Is that clear?"

The collective group answered, "Yes."

The judge sat down at his desk and picked up a sheet of paper. "This list of jurors that are being seated in the courtroom are average citizens. I will ask the standard questions. Then I'll ask prospective jurors if they are unable for health reasons or personal knowledge of the case to be seated as a juror in this case. I suspect that will take the rest of the day. I hope to seat the first group of jurors before we leave today. Tomorrow I will hear the defense and the prosecution make their selections and strikes of the jury candidates. I trust that should take no more than one day. That puts us to Wednesday. I expect if all goes well the next two days you can begin with your opening statements. Unless there are any objections, we'll plan to begin with opening statements on Wednesday. Any objections?"

The group responded, "No your honor."

Finally the judge looked up at Dobbs, "I assume Mr. Dobbs, you will be handling the entire case for the government?"

Dobbs answered, "Your honor…Ms. Lisa Williams will be handling parts of the government's opening case. She has an expertise as to how grant projects are applied for and funded by universities. She will lay the ground work for the jurors to understand the grant process. I will be handling the rest of the proceedings."

The judge looked at Downing, "I'm assuming you are the lead for Doctor Mathenson?"

"Yes your honor, I will be handling the embezzlement charges. Mr. Morgan will handle parts of the case that may arise that support our client's case."

Thompson looked over his glasses at him and said, "That may be very clear, at least to you Mr. Downing, but I am not sure what that means. Try and be a little clearer with your answers once we are in the courtroom. I may be older, so I trust you will not leave me wondering what you mean with your answers. If I don't understand, I am sure the jury will not either. Is that clear?"

He then looked at Ms. Williams. "You can speak, can't you?"

"Yes, your honor."

"Good, you have a full strong voice. I hate trying to hear what lawyers are saying. I trust that will not be a problem?"

She responded, "No your honor."

The judge looked at each attorney and said, "Once this begins, that courtroom is my domain. My world to make decisions. I am not a person who likes empty statements and long rambling storied questions. If you want to ask a question of a witness, ask the damned question. I do not like courtroom theatrics."

Before anyone could answer he said, "All right you may return to the courtroom. We will begin in fifteen minutes. We have a great deal of ground to cover seating a jury."

As Charlie and Downing were making their way back to the courtroom, Downing asked. "Is that typical of a federal judge to take the attorneys into their office before the trial begins?"

Charlie thought for a moment. "It happens from time to time. Judges like Thompson have seen it all over the years. He is well respected. That is one of the reasons they bring him out of retirement for cases. He doesn't have to prove anything. This gives him the ability to run his court the way he wants to. He knows where problems can develop. If he points out what he will tolerate and what he won't, it eliminates conflicts in the courtroom in front of the jury. It's kinda' like the referees in a football game. They bring the captains together before the game and introduce themselves. They then explain to the captains their rules and what to do if there is a problem during the game. You just got your pregame warning from Judge Thompson. He knows me and he knows Dobbs, but you and Ms. Williams are unknown. He just put both of you on notice and didn't have to do it in front of the jury. This case is high profile in the public eye. If he can prevent problems, it serves everyone in the course of the trial. Judge Thompson hates after several days of testimony, the jury cannot make a clear decision. If he has to declare a mistrial, he feels it is because the evidence was not clearly given to the jury. And if his court's decision is overturned at a later date because of some error during the trial...that also reflects on the judge."

Downing then asked, "Did you expect the gag order in this case?"

"Yes, I did. You get a case like this where the explanation of the evidence is so important, the judge wants to keep those explanations inside the court. If you get attorneys and clients talking to the media, it can change things dramatically. One of the lawyers or the person on trial could make a passing statement to a reporter. That evening it runs on the news. Jurors are people, they watch television even when they are instructed not to. Six o'clock news comes on and they see the person on trial say something, one of the jurors hears it, all the jurors will know it by the time they have deliberated. Well, this way it should all stay inside the courtroom. At least, that is the judge's plan."

As they returned to the courtroom the prospective jurors had been brought in. As they walked, Charlie said to Downing, "This will be your first time in the jury selection process in federal court. These people are pooled from all around the district. Some have come a long way to be here. I would bet the biggest part of them would rather be somewhere else. Most understand their obligation and will do what they have been called here to do. Look for ones that slouch in their seats, keep looking at their watches or around the room. Some may even tend to fall asleep. These are people who are not good jurors. While the judge is going over his instruction and questions to the group, watch faces. People reveal a lot about their feelings. Whenever you get to speak with the juror you can remember those people. You see someone who rolls their eyes. They most likely have already judged you. Remember those jurors not only judge the person on trial, they judge you. Your job at this point is to judge them and pick the ones we'll want to strike. Pay close attention. Your client's guilt or innocence depends on your selection at this point."

As Downing sat down at the defense table, he leaned forward to Charlie, "I'm not bad at judging people, but I will also glance at you if I have a doubt. If you see something I don't, close your eyes?"

Charlie half chuckled, "No problem, but if I keep them closed to long, I may have fallen asleep."

Elizabeth perked up at this point, "Don't worry, I'll keep both of you awake. I'll give you a little kick under the table."

The door in the front of the courtroom opened and the bailiff called out, "All rise. The honorable Judge Harold Thompson presiding."

The room was silent as Judge Thompson entered and sat down. Just as quickly he instructed the courtroom to please be seated. "All right ladies and gentleman, as you know you have been summoned to be a juror in this case. Is there anyone who has not filled out the questionnaire given to you regarding your address and miles traveled?" No one indicated a response to his question.

He then instructed the court clerk to take roll call of those in the jury pool. The roll call revealed all that had been summoned were in attendance.

Over the next two hours the judge asked the standard questions of prospective jurors. Occasionally a juror would hold up their hand indicating the question had raised an issue for them. Each time Thompson would have them approach his bench and speak with the individual to resolve the issue. Downing and Dobbs listened as the judge made decisions concerning the person. The court recorder kept track of the conversations.

After the judge completed his questions and instructions, he looked at the clock on the back wall of the courtroom. "Alright, it's one o'clock. We will recess until two-thirty. At that time the clerk will select the names of sixteen jurors and they will be seated. The first twelve names will be seated in the jury box. The other four alternates will sit in the gallery. At that point counsel for the prosecution and the defense will speak with each of those selected. I expect this will take the rest of the day and most of tomorrow. If you have not been selected by the end of the day, that does not mean you will not return tomorrow. Not until the selection is complete will I dismiss those not selected. Is that very clear?" No one said anything, but a slight audible sigh could be heard from within the gathering. With his instructions given, Thompson dismissed the courtroom.

Chapter 21
The Eyes of a Juror

Charlie and Downing worked through the recess going over their questions to the jury. Charlie knew from years of experience dealing with jurors, it's best to be politically correct in your questions. Charlie always found what worked best for him was to have the judge ask questions of the jurors, then he could watch their faces and eyes. He believed, through all his years practicing law, the eyes were the most revealing. That was followed by a person's body language. People being questioned sometimes try to hide something from the lawyers. Much like a poker player, a good lawyer needs to be able to read a person's tells when they are being questioned. Charlie knew everybody had their own tells, twitch of the eye, looking away, or rubbing hands together. Movements of the eye and blinking are involuntary and controlled by the body's subconscious mind. Charlie never figured himself to be a poker player, but he could read a group of people who figured they had nothing to hide. They were not on trial. They were there to judge someone else.

By twenty-five after two, all the jurors and the attorneys were back in their seats. Precisely at two-thirty Judge Thompson returned to the courtroom. As he sat down, he instructed the bailiff to gather the questions from the attorneys. He looked at the jury box making sure all the jurors were there, and then he looked out over the people seated in the gallery and did a silent count to himself. By then the bailiff had returned with the lawyer's questions.

He sat for a few minutes looking them over. Finally after five minutes of reading and making some notes of his own he turned to the jury. "Ladies and gentlemen, the prosecution and the defense have prepared questions they would like to have me ask. All that I need from each of you for responses is show of hands or a nod of the head, yes or no. If for any reason you feel it is necessary for you to explain an answer, I will have you approach the bench along with the attorneys. Are my instructions clear?"

Everyone sat quietly and nodded they understood the judge's instructions.

"All right, let's begin. How many of you have seen, read or in some way heard about this case?" As he finished he looked at those yet unpicked sitting in the courtroom and said, "I want you also to answer the questions. It is possible if for some reason one of the selected jurors is dismissed from the potential jury, one of you will be selected. Do you understand?" They indicated they did.

"All right, how many of you have read, seen or in some way heard about this case?" Most every hand in the room went up. The judge surveyed the response and glanced at the lawyers.

"Do you understand you have been ordered to this courtroom to render a decision in a case where the defendant has been charged with two counts of embezzlement, one count of wire-fraud and one count of misuse of federal funds?" The perspective jurors seemed to be a bit confused as to the judge's questions.

The judge surveyed the room and then called the attorneys to his bench. He covered his microphone. "Gentleman, I would assume that is the answers you had expected?"

Downing spoke first, "Yes, your honor that's about what I expected. Some of these people listened to the whole news story and why a trial was being held. Some of the others only heard about the secondary aspects of the case. I don't care how many people we ask, the responses most likely would be about the same. They were truthful in their answers. I can work with that."

Thompson looked at Dobbs. "Well counselor, what are your thoughts before we proceed?"

"The government would hope the majority of the people understand the charges in this case are filed under Section 18 U.S. Code 665, the embezzlement section. But, I am realistic enough to know, other than those of us in the legal profession, these people do not understand the statutes. It is my job to give them some understanding of the law and the charges in this case."

Thompson leaned forward, "Mr. Dobbs, you know the question I was asking. Do you have a problem with the answer that some of the people see this as an embezzlement case and others see it as being about…well I guess I'll say the science of her work?"

Dobbs thought for just a second, "No, your honor, we have no problem with the answers given by the jury so far. But I will stipulate, I hope this does not degrade into something completely foreign to the embezzlement case." He looked at Downing as he spoke the words.

Downing looked back at him, "I am prepared to concede nothing. The evidence will take us, where it takes us, I suggest we get about selecting a panel of jurors."

Judge Thompson with a bit of a grin looked at Dobbs, "Anything else from the prosecution at this point?" Dobbs frowned a bit and answered, "No your honor."

After they returned to their tables, Judge Thompson continued his questions. "Do any of you feel that you could not render a decision in a case where possibility science may play a part in the evidence?"

One man in the jury box raised his hand. The judge motioned for the person in the jury box along with the attorneys to approach the bench. Once again he covered the microphone. He looked down at the man and asked. "You have a concern?"

"Yes, your honor, I do. I do not know for sure what this case is going to be about, but I have heard the person on trial has created some kind of life using science. I think that is impossible. Only God has the ability and right to create life. And if she's using our government's money to do this kind of research, well I think you all should be on trial here if you're asking me."

Thompson looked at the two attorneys. Dobbs said, "I move to strike this juror on grounds, he may not be able to render an impartial judgement for either party in this case."

Downing looked at the judge and shook his head in agreement. After both indicated they wanted to strike him, the judge looked at the man. "Sir, you are excused from the jury, but you will wait in the gallery with the rest of the potential jury members. Is that clear?" He nodded in agreement.

The judge asked the clerk to call another name. In a few seconds the name of a woman in her mid-fifty's was called and she took her place in the jury box.

The judge looked at her. "You heard my question, did you respond that you could not make a decision in this case?

She answered, "No, your honor, I will have no problem in rendering a decision based on the evidence."

The judge began once again asking his general questions. One other juror was removed and replaced. Each time he made notes of the changes and why. After the last juror was replaced he looked up at the clock. It was going on four-thirty.

"The hour grows late and I have more questions. By my estimation, we will complete the jury selection in another couple hours. That is if we don't run into any problems. I have decided to continue questioning jurors and complete the process this evening. At that point, I will release those not selected. Do I hear any objections? Do you all want to continue until we are finished with the jury selection tonight?"

With a mutters of "yes" from the room, the judge continued. By quarter to eight he had completed his list of questions for prospective jurors. He motioned for the attorneys to approach the bench.

Thompson leaned over, "Do you gentlemen have anything else to add to this process before I declare we have impaneled a jury in this case?"

Both Dobbs and Downing answered, "No, your honor." With that he motioned for them to return to their tables.

"I want to thank you ladies and gentlemen who sat through this long day and were not selected. Make sure you have given all your information to the clerk. He then looked at the empaneled jury, five men and seven women with one man and three women as the alternates. Those of you who have been selected to hear the case will return tomorrow at one o'clock for opening remarks. I would caution you not to discuss this case among yourselves or anyone outside this courtroom. Especially the press. I have issued a gag order in this case involving all that are part of the case. Rooms have been booked for those of you who live out of town and indicated you need a place to stay. Check with the clerk before leaving. Again thank you for your patience, and remember, speak to no one about this case. Court is dismissed until one o'clock tomorrow."

Charlie turned to Downing, "Very good so far. I think our jury make up is as good as we could have hoped. Questions about science that touched on religion did not get any obvious reaction. Hopefully they will be open to hearing all the facts before rendering a decision."

Tucker stepped up and shook Downing's hand, "Young man, you seemed to be a knowing your stuff. Have you prepared your opening statement?"

Downing half smiled and said, "Pretty much so, Charlie and I worked on it for the last couple of days. We both feel satisfied with the opening facts I will tell the jury. Charlie explained how I have to give the jury credit. We have to believe they are people who are willing to listen to all the facts before making a decision. Too much lawyer talk tends to grow old if you are saying nothing to them. Those sixteen people know why they are here and they are counting on me knowing what my job is. I can work with that."

Jake stepped up, "I don't know about you all, but I am hungry, no supper break and my stomach knows it. The bailiff told me there is a place just down the street where they sell craft beer and have great steaks. Who's coming with me?"

Downing and Angus quickly said yes. Elizabeth said "If you don't mind, I'm going to call my son and check and see how the family is doing. You all go ahead. I'll order a pizza and have it delivered to my room."

They then looked at Charlie who had said nothing. As he closed his brief case he said, "You all go ahead, I need some time to think. If I remember, there is a nice place down by the river. I'll walk down there and watch the sun set. Maybe it will remind me of my fishing spot in the mountains."

In a few minutes the courtroom had cleared out and only the security people were left. Upon returning to the hotel, the group disbanded and Charlie returned to his room, removed his suit and put on his old comfortable every day clothes. He searched around his suitcase and soon found what he was looking for. A silver case with his cigars inside. He removed one and stuck it into his shirt pocket and found his way to the front door. As he stepped out onto the street, he was surprised how quiet it had become in the last hour. Down the street he could see the rivers water reflecting the last rays of the day's sun. He decided to walk towards the river. As he walked he tried to remember the last time he had been in Wheeling. It had been several years and much had changed.

It took him ten minutes to make his way to an empty bench near the river. He watched as coal barges passed by heading north and a few minutes' later barges loaded with coal passed heading south. Charlie wondered to himself, "could this be a very efficient way of transporting coal? Couldn't the upriver coal supplies service the upper river customers and downriver the same?"

After just sitting watching coal barges pass by, he removed the cigar from his pocket and rolled it in his lips, wetting its end. Then he bit off a small piece and spit it into a nearby trash can. Next, he flicked the lighter and held the cigars end in the flame. Slowly he began drawing in the smoke. He held it for a moment, then he broke into a thirty second coughing fit.

From behind came a voice, "You know Charlie, those things will kill you."

He recognized the voice. "Harrison, you should not be here. Someone sees us talking and they will think we are collaborating on the case."

Harrison sat down beside him. "Well Charlie, I'll promise you, I will not say anything about the case, if you promise not to talk about my political aspirations. How's that? Will that put your mind to rest?"

Charlie chuckled as he ignored Harrison's conditions. "How's life been since we last talked? I heard you are on the fast track under the new president to be nominated for the Supreme Court. They say there is going to be an opening later this year. Some say you will take the bench, having never been a judge in your career. Straight from the prosecution table to the highest seat in the land. The power of political correctness. Do you know where the term political correctness supposedly came from? I heard that Truman told MacArthur to be politically correct during the surrender of Japan. MacArthur sent a message asking what political correctness meant. The story goes he explained the term to be similar to trying to pick up a dog turd by the clean end." Both men chuckled at Charlie's story.

Harrison did not answer Charlie's question, but instead said, "I thought you quit smoking years ago. In fact, the last time we worked together you were chewing nicotine gum to help you quit."

"I did for several years. Then Rose and Timmy were killed…and somehow I didn't think I had a reason to stay healthy. Still, when I smoke on rare occasions, it's just not the same. I think I want to remind myself that I had an agreement with my wife and I remember how on occasions I would go on the back porch and sneak one. She pretended she didn't notice and I pretended she didn't know. When I smoke, I think Rose is still watching and giving me that look."

Charlie looked at Harrison, "But you know what? It seems with each passing year I can see her face less clearly. I miss my wife and my son. He would have been a young man now."

Harrison said, "I am truly sorry about your wife and son. I heard you made sure the man who did it will never see the light of day again?"

Charlie took a small drag on the cigar and blew it toward the setting sun, "That's what they say. But I'm guessing someday his family will find a judge who will reverse his sentence to time served. You know how judges can be in an election year. Takes friends with money to run for office."

Harrison half chuckled at Charlie's comments. Neither man spoke as the sun touched the far mountain.

"Harrison, can I ask you a question? You don't have to answer if you don't want to. How come you took this case? You had to know some of this cases background beforehand?"

Harrison thought for a minute before answering, "I guess the answer is not so much about the case, but about me. The truth is, I was asked to take on this case, figuring it would be resolved ahead of time. The higher ups believed she would not want to go to trial. The university was not really that eager to have her department drug through the courts. She was offered a settlement and she refused. By then I was already involved. If I recused myself it might raise questions. If I assigned it to one of my assistants and they lost, then that would reflect back to me. In hindsight I should have never agreed to look into this case in the first place. I should have had another district prosecutor handle it, especially after I learned what was behind the case. Charlie, you had to prosecute a few cases in your times that started out to be fairly simple. Then something is uncovered and things get all screwed up?"

Charlie thought for a moment and chuckled as he remembered some of his own history. "Yea, I guess I had a couple of cases that ended up like that, but that was a long time ago."

Harrison sat back and relaxed, "Besides Charlie, It's not personal with me. It's just a job. One more simple embezzlement case."

Charlie spoke up, "It's not going to be just another embezzlement case, and what the hell does it mean, not personal. Every time we go into that courtroom we are expected to make a difference. When I was in your position I was supposed be the keeper of the law. Protect the American way of life with justice for all, especially the guilty."

"Truth is, Charlie, you were a lawyer's lawyer. You had ethics even when you were prosecuting people. You got convictions without having to slant the evidence to sway a jury. You were good at being a prosecutor. As to me, well I feel like a television lawyer sometimes. I am good and I know it. I like the cameras and I am ambitious. In spite of all that, you have to admit, I know the law."

Charlie watched the sun disappear as he said, "Well, I will admit, you do look good on camera, but you'll not be in front of one until this is over."

Charlie nudged Harrison, "I think you got a lot riding on this one."

Harrison paused for a moment before saying, "I learned a long time ago that there is more than one way to pick up a dog turd by the clean end. That's why I look good on camera." They both chuckled and said nothing as the daylight faded from the sky.

Chapter 22
Opening Day

Charlie spent the early morning working with Downing. He had him running through his opening statements to the jury. Charlie wanted him to know his words until they became second nature to him. He emphasized the fact the jury sees Elizabeth through Downing's presentation and his actions each time he stands before them. If he loses credibility with them, it reflects on Elizabeth's case.

As they were preparing to leave a small conference room in the courthouse, Tucker asked. "I know that some of you may not be Bible carrying Christians, but I would like to ask the man upstairs for a little help in today's proceedings if you don't mind."

Jake spoke, "Fine by me, but can you extend the prayer to cover the whole trial? No sense spending a good prayer on just today. We'll be here for a while, I'm guessing."

Everyone nodded in agreement to both men's request.

Tucker began, "Heavenly Father, we come before you on this day to ask that you give strength and wisdom to Elizabeth's case before the court today. Guide James and Charlie in her defense. We ask that you grant the jury understanding along with patience while listening to the words of her defense. And Lord, I know there are those people who are saying Doctor Mathenson has violated your laws and should be punished. We ask in your eternal wisdom that you understand her intents are to ease mankind's pain and suffering. Her intent was not to do the work of creation. In your name we pray, Amen."

Charlie picked up his briefcase as he said, "Well Tucker, I'll give you this, you cover about everyone with those words. Let's hope he was listening."

Tucker smiled at Charlie, "He was my son… he was."

As they stepped into the hallway, Marshals Hickman and Harris were waiting. Samantha looked at Charlie. "All right, are you all ready to run the gauntlet of news people and crazies in the front hallway?"

Charlie said, "Let's go. Concentrate on getting Elizabeth into the courtroom. We'll be fine. She's the important one."

As they turned to go down the hallway, building security was working to keep the doors and hallway open. Cameras were pointed toward them as they approached. Countless flashes from cameras nearly blinded Charlie as he looked toward the crowd. A roar of questions were being called out to Elizabeth as she kept her head down trying to avoid the blinding flashes.

Suddenly a woman with a tee shirt on that said "God is Love" burst through the security perimeter and grabbed Elizabeth by her arm. She cried out, "What kind of a doctor are you, creating the offspring of the devil? God will punish you." As security pulled her away she looked at Tucker. "And what kind of religious man are you. Satan send you to protect this whore of the Devil?"

For a moment confusion broke out in the hallway. Charlie and Jake pushed themselves in front of Elizabeth to add a layer of protection. At the same time they were pushed by the crowd toward the courtroom doors. Marshals Hickman and Harris kept their bodies between the crowd and Elizabeth as they finally entered the courtroom. In a moment, Charlie, Jake and Tucker followed them in.

"Is it going to be that crazy every day?" asked Downing.

Samantha answered, "No, we'll not go through that again. We'll enter from the back hallway near the judge's chambers. Building security was supposed to have that under control. Obviously they didn't."

For the next few moments the group regained their organization as they moved toward the defense table. Five seats in the row directly behind them were reserved for Jake and Tucker. There was also a seat reserved for Elizabeth's son.

Elizabeth spoke quietly for a moment to her son before giving him a hug. Then Elizabeth turned to her defense team as she said, "This is my son Robert Mathenson. He's a doctor in residence at the university hospital. He can't be here every day, but he wanted to be here for the opening."

Everyone shook hands and exchanged greetings. When Robert took Charlie's hand, he held it for a moment. "I want to thank you for helping my mom. All through her life she has only concentrated on finding answers to medical questions. She never took the time to look around the world or take stock in what her dedication was costing. And I don't mean in terms of money, but personal cost. She is hard headed and she's stubborn, but she is none of the things the media is calling her. I can't understand how they have demonized her so much. You know that Jedidiah Cain, the pastor of the Cathedral in the Sky, is saying she is the bringer of the Apocalypse. Her science will destroy all mankind. He has been on television crusading against her and science. He is telling people his words are those of God. And God has chosen him to deliver judgment on my mother and any that follow her. It has gotten downright ugly. I have moved my family to protect them from this madness."

The young man squeezed Charlie's hand, "I know you and these other gentleman will do all you can to help her."

Charlie nodded, indicating he understood Robert's words. He then said to the young man. "You understand we have a difficult case here. And you also need to understand your mother has dismissed offers by the prosecution to plea bargain this case away. She has refused to do that. She has not even acknowledged missing research documents, along with a cryogenic chamber they alleged she has removed from her research lab. If you can, it would go a long way if you could convince her to return those items. I suspect you have some personal knowledge of that canister. As to what's happening out there in the world concerning this case, I can't worry myself with that."

Charlie pointed toward the jury box, "Those twelve people who will sit in those chairs will decide your mother's fate. I would like to tell you they have not seen or know what is happening outside those doors, but I can't. Our only hope is they are good conscientious people who will listen to the judge's orders and render a decision only on what happens in here, not out there. But, you will have to admit, that will be difficult for any of us to do."

The young man shook his head, indicating he understood Charlie's words. He then went on to shake Downing's hand, along with Tuckers.

As Robert was speaking with his mother, Jake nudged Charlie and pointed toward the seating in the courtroom. "What the hell? Aren't they going to let anyone into the courtroom besides those religious guys in the back row and news people?"

Charlie had been so caught up in thinking about the beginning of the trial, he had failed to notice the room was almost empty. He looked at Dobbs as he shrugged his shoulders and mouthed quietly. "Where is everybody?"

Dobbs motioned for him to meet him in the center of the walkway. "There are a lot of people waiting, especially the news people. The judge has decided to let the media pick who they wanted to be in the room. He limited them to ten, representing the media. Those who are in here today are expected to switch off tomorrow. They are also expected to share their notes with other reporters. Kind of a media pool at least for today. The second courtroom should remedy the problem tomorrow"

Charlie tipped his head toward the two separate groups of religious men. There looked to be several different faiths in the group of twenty. Charlie asked, "What about them? They are not with the news. How did they get in so soon?"

Dobbs half grinned, "You remember last night when we talked about political correctness and how you handle it? Well, I was asked, no, that's not correct. I was told to pick up that dog turd by the clean end and allow them preferential treatment in being seated in the room. I had no choice in the matter. And before you ask, the four guys who look like they work for some secret government agency, well they are from the State Department. Hell, they won't even tell me why they are here. If you have anyone you want to make sure gets in each day, just tell the head bailiff. The people outside had to put their names on a list to get in. The judge is only going to allow those who signed up to enter and in the order they signed up. When the seats are full, security has been instructed to no longer let anyone in. I think I mentioned, tomorrow they are going to run closed circuit television into a second courtroom. They did not see this coming. Truth is, neither did I."

Charlie looked at the collection of reporters and clergy men in the room as he said to Dobbs. "And you think we are going to keep this from spinning out of control?"

Charlie returned to the table and sat down removing his file from his briefcase. He sat for a moment organizing his thoughts. As he did he noticed that Elizabeth looked really upset. He leaned over to her and said, "You need to relax. We have not even started. All this around us is just noise. You have to put it out of your mind and remember why we are here. And remember, you can end this now if you take the government's deal. But when the judge comes in and we begin, it will become much harder to stop this. Do you understand?"

Elizabeth looked back at her son and smiled at him. Then she looked back at Charlie, "No, I am all right. I am not guilty and I want it proved in front of the world. I did not steal any money."

Charlie looked her coldly in the eyes, 'Elizabeth, all this is not about the money as I once believed. It is about the work you have done and the answers you will not give. Don't convince yourself this is just about the money. Hell, I don't even believe it's about the money now myself."

He paused for a moment before asking her, "I will ask one more time, is this what you want to do?"

She composed herself as she looked back at him. "This is what I want to do. I knew this is where I would end up. I just didn't anticipate everything else that is going on."

About then the doors opened to the courtroom and security began allowing people into the seating area. A third security guard counted those entering. When the count reached the point where the seating would be full, he signaled to the guards at the doors to prevent any more people from entering. Those outside were clearly upset they were not going to be allowed access to the room as the doors were closed.

As the courtroom settled down, a door near the front was opened and the jury filed into their seats. The room went quiet waiting for the judge to enter.

For the next five minutes the room waited. Finally the bailiff opened the door to the back chamber and the judge quickly entered the room. The bailiff was caught off guard by the judge's quick entrance. He abruptly announced to the courtroom. "All rise, the Honorable Judge Harold Thompson presiding."

He sat down and as quickly said, "You may be seated." He arranged his desk and adjusted his glasses. After a few moments he looked out over the courtroom. The judge could hear the commotion in the hallway.

He called to the guard near the back. "What's that noise in the outer hallway? I want that stopped NOW!"

One guard opened the door and slipped out of the room. The other guard said, "Your honor, those are people who were not allowed in because of the large number who wanted to attend this trial. I understand tomorrow close circuit television will be tied into courtroom "B". That should eliminate any problems tomorrow."

The judge thought for a minute. "Well that may take care of tomorrow, but I don't want to hear that noise today. Either they wait quietly or they need to go into the lower lobby. I don't care which, but I want it done now. Is that clear?"

"Yes, your honor." With those instructions the second guard disappeared through the double doors into the hallway.

Judge Thompson stared toward the doors until he could tell it was beginning to quiet down.

"All right, we are here today to begin the trial of Doctor Elizabeth Mathenson. The federal government has charged Doctor Mathenson under Section 18 U.S. Code 665. Theft or embezzlement from employment and training funds; improper inducement; obstruction of investigations. The indictment alleges that she directly, or by intent, caused monies directed to research programs governed by university guidelines to not have been used according to original grant purposes"

The judge removed his glasses and looked at the jury. "Do you understand what I just said?"

About half the people slowly raise their hands in response to his question.

Judge Thompson leaned onto his forearms on his desk as he spoke. "In simple terms, it means the government is saying Doctor Mathenson stole or misappropriated funds entrusted to her and did not belong to her. She is alleged to have stolen or misused funding belonging to the research department of her employer, the university. Does that help with your understanding?"

This time everyone raised their hands indicating they understood his explanation.

Judge Thompson then directed his attention to the attorneys in the case. "At this point before we begin with opening statements, are there any motions or recommendation that need to be brought before the bench?"

Harrison Dobbs stood and said, "No, your honor. The government is prepared to proceed."

Judge Thompson then looked at James Downing. He stood and faced the judge. For the first time Charlie could detect a bit of nervousness to his voice as he spoke. "No your honor."

Downing paused and cleared his throat, allowing his voice to steady before starting again. "The defense is prepared to proceed at this point your Honor."

Thompson sat back in his chair as he looked at the prosecution's table. "Mr. Dobbs, I believe you are up first."

Dobbs stood and buttoned his coat, "Thank you your honor."

He walked to an area just in front of the prosecutions table, facing the jury before he began. He at first stayed back from crowding the jury box. "Ladies and gentleman, good afternoon. You have been selected to hear evidence in a case of embezzlement. The judge has given you an explanation of the meaning of the word and the intent of the law. If you were here to sit in judgment of an everyday theft of money by robbery, it would be an easy task for me to prove to you that case. I would simply show you the evidence against the guilty party. Then I would tell you how that person committed the robbery. Next, I would show you the evidence used to arrest the person. In some cases I could show you video tape of the accused person taking the money. With the latest forensic evidence, I could use finger prints on a gun to prove they were the one who committed the crime. I could even produce an eyewitness that could testify about the person on trial was the same person they saw committing the crime. I could even have them point their finger directly at the person in this courtroom, proving to you they were the person who committed the crime. For you, the jury, it would be easy to hear the evidence, see the proof and convict the accused person. But, this is not one of those cases. The crime of embezzlement is about removing financial funds or material items by hidden means. Embezzlement crimes typically occur over a period of time. Many times when someone is caught, it takes a good while for the case to be brought to trial. The theft of funds could go back years. And when it is revealed who the person is that took the money or material items, everyone is often shocked. It often turns out to be someone they saw every day doing their job. A trusted co-worker or even a friend. But that person, for one reason or another, had been illegally removing property or money owned by the company for whom they worked. It may be the government. Or even a church. Or in this case it is a place of higher education, the university. A place the person accused of this crime worked for many years. She was a trusted and revered doctor, known around the world for her research into new forms of medicine. Such is the case you have before you today."

By this point Dobbs had positioned himself in front of the jury box. Slowly he looked at each juror's face. Dobbs was a seasoned lawyer and knew not to talk down to people, but to speak slowly and precisely, choosing his words. As he drew closer to the jury he knew they were at ease with his presentation. He sensed they were listening and understanding his words clearly. He hoped they understood the points of law that he had spoken of and how they showed Elizabeth guilty of embezzlement.

Dobbs began, "Over the next few days I will have people who write grants explain how they work. I will have an accountant explain how the money is handled once it is brought into the program. And I will have financial controllers from the program explain their jobs and how they worked with Doctor Mathenson over the years. I will also bring to the witness stand some of her co-workers. They will explain that Doctor Mathenson's research began when she was a brilliant young doctor. They will then go on to explain how her approach to her work changed over the years. How she became a doctor crusading to spend the money for scientific goals not part of the original grant money guidelines. They will tell you how she managed to fool everyone by diverting money, even to foreign governments, to further her private quest for science."

As he spoke he walked slowly in front of the jury, occasionally holding his hand toward them with his palm up as if he was offering them something. A gesture of friendship and offering. Dobbs was good at drawing people to his voice and explaining the case.

Finally he moved back to the center of the room. He stood for a moment looking at the jury and offering a slight smile. "You all have been asked to sit in judgment of a fellow human being. To decide the fate of someone you don't even know. The defense may tell you this is a victimless crime. Everyone steals from the government. No one was hurt. In fact the money was given to her to do research. Don't be misled. When anyone, you, or even me, is given the trust of property that belongs to someone else and you don't use it for the purpose intended, you have broken that trust. Trust, remember that word. Trust is all that keeps our world from falling into madness. Just think about your own lives and how many people you trust to do the right thing, even when no one is looking. Doctor Elizabeth

Mathenson broke that trust. And I am confident when you have heard all the evidence, you will determine she is guilty as charged. Thank you."

Dobbs stood straight and tall throughout his entire presentation and as he returned to his table. He sat down and adjusted his coat. While this was going on the jury watched his every move. He had clearly touched each juror and gained their confidence with his opening remarks.

Judge Thompson made a few notes as he looked down at his desk. He sat up and cleared is throat. "Mr. Downing, are you prepared with your opening statement to the court?"

Downing stood as he answered the judge's question, "Yes, your honor, we are."

Downing stood and picked up a book that sat on his table. He seemed to weigh it in his hand as he walked toward the jury box. "When I first went into law and began researching cases that our professors had given us to study, I wondered how could I carry so many heavy books."

Again he weigh the book as he looked at the jury and he gave them a smile. "Within this book is Section 18 of the U.S. Code 665. Some may say it has everything you would ever want to know about the federal laws governing embezzlement. Judge Thompson has given you his fine explanation of the law. A very good explanation. He read it from the legal definition. And then to help you understand it a little better, he broke it down into terms we all use every day. Mr. Dobbs, the very experienced prosecutor in this case, explained the reality of the word embezzlement to you. How did he define it? Someone steals from someone who trusted them to oversee something of value. He explained how over time a person might take what is not legally theirs right under the noses of other people. A crime that you or me would never commit. People sometimes reassure themselves there is no way anyone would take anything from a trusted employer or friend. But, we know it happens. For instance, let me give you an example of what some may technically call embezzlement. Let's say your job is to purchase light bulbs for your place of work. You control the money budgeted to buy light bulbs when needed. Now, let's say along the way in the future the light bulb over your boss's desk went bad. It just quit working. When you contact building maintenance, they tell you the light fixture needs to be replaced and you will have to write a work order to receive funds to purchase a new one. Repair money comes out of a different account, you want to expedite the process, so you purchase the fixture using money from the miscellaneous funds. Those funds are used to purchase a new light fixture along with new bulbs. Problem solved and your boss is happy."

Downing raised his hand quickly over his head and said, "Let there be light and there was light. Now, did the spending of the money from the miscellaneous fund on the light fixture constitute embezzlement on the part of the person who was trusted to buy only light bulbs with that money? Or did that person use their discretionary judgement to solve a problem? You can see the answer is not always black and white. We will prove to you that Doctor Mathenson acted not with intent of misuse of funds, but the evidence will prove she acted in the best interest of the programs goals. Goals that changed as advancements in science were discovered by her research and others."

Downing held up the heavy code book and said, "Contained in this book are hundreds, no thousands of the reference terms, whereas, in the case of, or discretionary judgements." Legal definitions that try and answer many different variations and interpretations of the law. I can tell you I have spent many a long night weighing the evidence in the cases and trying to understand the intent of the law as written in these books. It's not always easy to do. And now you have been asked to do what I spent nights trying to understand and learn the complexities of the law. If that is not enough, you have to review settlements from other courts that further define the laws contained in this book."

Downing looked at the jury and said, "If you don't mind, I am going to lay this book back on the table. I didn't realize the law would be so heavy." Several in the jury box gave a slight smile at his words.

After he laid it down, he turned and walked back to the jury and stood looking a bit humbled. "As you can tell I am not a seasoned attorney. In fact this is the first case I have stood before a jury in a real courtroom. Do I tell you this so you may feel I don't know my job? No, I want you to know I passed the bar and hope someday to be an experienced trial lawyer. But whether I have years of experience in courtroom trials or this being my first case, I will tell you this now without any hesitation. The prosecution is casting so broad a net with the charges against my client, they are hoping to confuse me. And even worse, they are hoping to confuse you in the process. What do years of trial experience give a seasoned attorney? It gives them the understanding of what influences a jury. For my part I will give you the law as it is written in those books." He walked quickly to the book and pointed at it. I'll tell you of the law and how it affects my client."

Downing paused and looked at Dobbs, as he sat watching his presentation. "The prosecutor told you a little earlier that I would try to convince you this is a victimless crime. How everyone steals from the government. Well let's be clear about this point. Steal or embezzling funds from an employer or friend is not a victimless crime. The funds Doctor Mathenson is accused of embezzling are your money. Each of you pay taxes to the government. You expect the government to make sure those funds are spent properly. So what does that mean in this case?

Every one of you has a stake in the outcome of this trial. It's your money we are talking about. Doctor Mathenson is not some phantom the government wants you to believe I am protecting in this trial. I am here to prove to each of you that Doctor Mathenson has not betrayed that trust that each of you gave to her. The evidence you will hear in this courtroom will prove Doctor Mathenson is not guilty. If the government, in its great wisdom, feels a person's guilt or innocents is judged entirely on the words written in that law book… that's just not the way it is."

He paused once again. "Now, as I close… I want you to remember my example of that light bulb. Did the person buying the light fixture in which the bulb was to be placed commit a crime? Or was it a judgment use of the money to accomplish the same task, restore the light over her boss's desk. I ask you on behalf of my client to look at the entire meaning of the law. But more than that, I want you to look beyond just words in that law book. I want you to understand the intent of the law in everyday application. Thank you."

Downing sat down at the table, and under his breath exhaled deeply. Charlie leaned behind Elizabeth and said in a hushed tone. "I think you deviated from the prepared script. But, you did very well."

Downing looked down at his hand and held it just off the table to where Charlie could see. It was still shaking just a bit.

Judge Thompson finished looking at some papers on his desk as he looked over his glasses to Harrison Dobbs. "Is the government prepared to call its first…"

From beyond the doors to the courtroom a commotion could be heard outside. Then loud voices shouted some unrecognizable words. Suddenly the doors were pushed open and a small crowd of people pushed into the room. They carried signs and were chanting, 'Doctor Mathenson is the anti-Christ. She must be destroyed!" The crowd of people overwhelmed security as they pushed their way in, chanting their message.

Judge Thompson slammed down his gavel and shouted to the bailiff, "Call down for more security. Hurry before this gets out of control." He then looked at the bailiff next to the jury box. "Get those people out of the room, now!"

With that the jury were quickly ushered out the back door into a secure area. Thompson stood looking at the unruly crowds in the back of the courtroom.

In a few minutes, sheriff deputies, city police officers and plain clothes marshals subdued the group of people. It looked to be about a dozen people that were taken into custody. Once they were under control, the head bailiff asked the judge what he wanted done with the unruly mob.

"I want them taken downstairs and processed. I will hold each of them in contempt. I will deal with them later. Now, get them out of here. Now!"

In a couple of minutes order was restored and the judge assessed the situation. In a moment he motioned to the attorneys. Downing and Dobbs quickly approached. Thompson looked at Charlie and motioned for him to also approach the bench.

He leaned forward and covered the microphone. I know we just got started, but I want things to stop and allow security time to make sure this does not happen again. The jury had to see what was happening before I sent them out. I think it would be in the best interest to call it a day, let security make arrangements and have building maintenance prepare courtroom "B". This got out of hand, I will not convene this court again until I am sure it can be done safely. This is a federal courthouse. I am going to find out what happened. Do either of the councils have a problem with my decision?"

Dobbs looked back at the now quiet courtroom. "I think that is a good idea. Are you going to bring the jury back in and explain what you intend to do?"

Thompson looked over at Charlie as he said, "I am inclined to let both lead councils make the joint motion upon bringing them back in. That way they'll understand this is a situation both parties want to resolve before we resume the trial." Again he looked at Charlie who tipped his head in an agreeing manner.

Judge Thompson motioned for the attorneys to return to their tables as he directed the bailiff to bring the jury back into the room. Once they were seated, both attorneys stood.

Dobbs began, "Your honor, in discussions with the defense team, we have concluded it would be best to recess for today and resume witness testimony in the morning. This would allow for measures to be taken to prevent any further disruption to the trial."

Thompson looked over at Downing. "Mr. Downing, I understand both you and the prosecution feel this would be your recommendation at this point?"

Downing looked at the jury as he said, 'Yes, your honor. We believe any further interruption today would be detrimental to the trial. A fresh start tomorrow would be the best for all in this situation."

Thompson looked at the jury, "We will resume at ten o'clock tomorrow morning. I will remind you that you are not to discuss any of the proceedings with anyone outside this courtroom. Is that understood?"

The jury nodded that they understood. He struck his gavel and dismissed the jury for the day. He quickly exited through the back door.

As the attorneys placed their files back into their briefcases, the bailiff asked if they would join the judge in his chambers.

As they entered the door of his office, they could hear him having a disagreeable conversation with someone on the phone.

"I don't give a damn what happened on the first level, you let those damned crazy people onto the second floor and they overwhelmed my court. You better make sure this never happens again, is that clear?" With that he slammed down the phone so hard it bounced off the receiver's base. He then picked it up and placed it in the base.

He looked up at the four attorneys, "I will entertain any motion for mistrial in the morning, if you feel this tainted the jury. Those people who were protesting knew what they were doing. A couple of them started a fight on the first level and security rushed to contain the problem. As soon as the protesters saw them heading that way, they stormed the courtroom. The captain of the security says he believes Reverend Cain's people orchestrated this mess. Well, they'll spend the night in jail waiting for me to impose a fine on them in the morning. That's about all I can do. They didn't damage any property or injure anyone. Best I can do is chew their asses and fine them. But I'll promise you this, if you choose to continue with the case tomorrow, it will not happen again. Now, go talk among yourselves and let me know at ten o'clock tomorrow before I bring in the jury what you have decided."

The attorneys could tell he was in no mood for conversation at the time. They knew the last thing in the world a judge wants is to lose control of his courtroom. Judge Thompson clearly did today.

As the attorneys stepped back into the courtroom, Dobbs looked at Downing, "Well, I'll let you make the call. After all, it is your client that is on trial. Do you think you want to start all over again or move along with the case tomorrow?"

Downing thought for a moment and then he motioned for Elizabeth to join them, "Doctor Mathenson, we are pondering asking for a mistrial because of the disruption so early in the case. I don't believe the judge will have a problem if we want to start this all over again with a new jury in a couple months. It is your choice. Do you want to forget today and resume the trial in the morning? Or do you want a new jury and trial?"

Without hesitation Elizabeth said, "I want to go forward. The longer this goes, the more is going to be said about the trial. You know damned well every one on that jury will go home tonight and watch it on television. At least those jurors were here and heard how crazy those people were. They each can judge for themselves today's events. Any future jury will see the report and form an opinion of the events based on a sixty second news report on television. No, we'll go forward. I trust the jurors will feel the same way."

Downing looked at Dobbs. "You heard the lady, we'll start again at ten. I hope there are no more surprises. I'm beginning to think I should revise my ideas of becoming a trial lawyer.

Dobbs half grinned as he said, "You'll do okay, we'll see you all in the morning."

Chapter 23

Whore of Babylon

Elizabeth's entire defense team, along with her son, gathered in Charlie's room to watch that evening's news. Not only did the local news stations carry the story, but all three major networks and a couple of cable channels picked it up. It wasn't surprising since they already had their satellite trucks in the area for the trial. The small riot near the courtroom doors just boosted the story to the top of the news cycle. Someone had positioned themselves to film the protest. Then they uploaded the segment to the news channels. The entire event had the appearance of being staged by someone who wanted to make the story even bigger. And whoever was there to film it had prior knowledge something was going to happen.

The major problem with the news stories, they featured Reverend Jedidiah Cain and he was being allowed to define the story outside the courtroom. Judge Thompson's gag order prevented anyone officially part of the trial from commenting about the case. But, Cain was not part of the legal proceedings and not bound by the gag order. With the gag order in effect, Cain was in his element in front of the cameras. He was the only one telling the story of how this case was about the biblical Apocalypse, as told in Revelation. He spoke directly to the television audience as if it was his show. He seemed to look those watching directly in the eyes as he proclaimed, "Doctor Mathenson's research had manipulated nature and was trying to create man in her own image." He went on to compare her to the Whore of Babylon and decreed her science could bring forth the Beast from the earth.

Cain proclaimed with arms outstretched for the cameras, "In the book of Genesis, God created the heavens and earth. And he said, this was good. Then he created man. Now, a WOMAN who God created from man has brought forth the spawn of the

Devil and she calls this abomination man. If the courts of justice in this land allow this terrible thing to happen, there will be such an uprising of the faithful, God himself will surely strike her from the earth and all those who follow her false religion, science."

Elizabeth looked at Charlie, "Where is he getting this from? The spawn of the Devil? And he never even mentioned the money."

Charlie took a deep breath. "He somehow knows about the DNA specimens and the cryogenic chamber of human eggs. He has not said it yet, but I can tell by his statements he knows. Someone has tipped him off and this very public trial will serve his purpose to do what he has always wanted, focus attention on himself and his ability to save mankind."

He looked Elizabeth directly in the eyes, "I told you, this is getting out of hand, but you can still end this. But, you now know the devil is out of the bag no matter what you do, from this point forward, your life will never be the same." He then looked at Elizabeth's son.

"Young man, I would guess you have a promising career as a doctor. And your mother tells me you have a family. This trial will spill over from the courtroom and into all our lives. Those news reporters out there only care about the headline-grabbing story. And the religious fanatics…I don't know for sure what they do care about, but your family is an easy target for them to hate. I use to say this case was only about the money, it is not about the money anymore. It is about people's beliefs. No, not about their beliefs… about their fears. And this sideshow preacher is going to make your mother the main attraction in his religious show. And he will in all likelihood drag your family into it. As of now he can't get to your mother, but if he discovers your family, he will get to them. His crusade has nothing to do with God or faith, it has to do with him stirring people's fears about their fate and pointing toward your mother and saying. "If you want to see the devil here on earth, there Satan is in the form of that woman and I alone can save you from this apocalypse."

Elizabeth went to Robert and put her arms around him. "Enough Charlie, you are scaring him!"

Robert placed his arms around his mother and looked at Charlie and the others. "Mom still thinks I am ten years old. I grew up knowing my mother was not like other moms. I didn't for many years understand what her goals were in life, but I know she has dedicated her life to making medicine better through her science. I once believed I wanted to have her dedication in my work. The truth is, I love being a doctor, but not to the point of ignoring my own family. I think I can be a good family man and balance being a good doctor at the same time. I also believe my mother's work will someday save thousands of lives. If she chooses to go forward with this trial, I will support her. I have already made arrangements to protect my family. I know in the beginning you told her this was about the money. I knew that as soon as the word got out, that would all change. But, I will have to admit it changed far more than I ever imagined."

Jake made a sound clearing his throat, "Well, what now? She is wanting to go forward and that's not going to be easy. I am not sure what witness we can put on the stand that can change the way the world is now looking at the doctor. At least the ones that have big opinions and even bigger mouths."

Tucker who had been sitting quietly listening to the conversation, said, "Forgive me, but I am no scholar of the law and what is to come. But I am guessing the prosecution is not going to go down the road toward those who disrupted the courtroom. I believe he will stay the course. He'll call witnesses that will make the government's case that Elizabeth misused the money. There is no way they will muddy up their case with anything else. Dobbs strikes me as a man who is ambitious, but not prone to twisting the facts of the case in order to get a conviction. He will raise questions about the money. There is only one person who can answer those questions and show the world she is not the devil."

He looked at Elizabeth, "I think you have to speak for yourself. Everyone in the courtroom and the outside world will be looking at you for answers they want to ask. First I think you have to explain where the money was spent. That's what the jury has been charged with making a decision on in this case. Your answers will help them with their decision, are you guilty or innocent concerning the charges? You are not on trial for the questions raised by your research, but there is no way around the fact everyone in the courtroom is wanting those answers. Did you create life? Let's all face it, that's the biggest question man has asked himself since the stories in Genesis. You can't sit in that courtroom and say nothing. You can't let opinions be formed from a man who tells the world he speaks for God. Reasonable people know the truth about him, but if he is the only voice in the emptiness of ignorance, they may begin to believe him. I believe there are two trials going on. One in the courtroom where twelve people will judge you. And the second is outside in the world. In here you have Charlie and James protecting your rights. Out there is only that self-righteous preacher speaking to the world jury."

He leaned down in front of Elizabeth, "They want to hear the truth from you."

Tucker looked around the room at the others, "I am sorry, but I felt I had to say my piece. I don't speak for God, but I do listen to him. He tells me to be truthful in life and have faith in people to do the right thing." With that said he sat down.

Downing looked at Charlie, "What do you think? This is way above my knowledge base. They never taught me anything like this in law school."

Charlie did not answer as he pulled the curtains and looked out on the street below. As he stared toward the passing people he said, "I would guess everyone down there has heard about this case. Some have formed an opinion and others thinks it's a story blown out of proportion by the liberal news. Others will believe it is the rock solid conservative truth."

He turned and looked at Tucker, "You are right. In the beginning I would have never thought about placing Elizabeth on the witness stand. And when it comes right down to it, she'll have to decide what to do."

Charlie knelt down beside her. "It will be at least a couple of days before you have to decide. Dobbs will put accountants, grant writers and most likely the controllers from the university on the witness stand to explain how the money was used. Then, we will call our witnesses. We have lined up other scientists to explain how funding was used in different parts of the research program. They'll explain your work was compartmentalized to keep aspects of the research from interfering with other departments. I believe their testimony can explain to a reasonable jury about your funding and how it was used for research, not embezzled as the government contends. But, when it comes to the beliefs of the jury that you can create life, only you can answer those questions. Those embezzlement charges will be forgotten by the world at the end of the news cycle. No one cares about that. But the question of your having found the secret to creation, that story will take on a life of its own and will be part of all our lives from here on out. We will support you and defend you with our best defense no matter what you decide. But, the decision is yours and yours alone. One more thing, Dobbs will get a chance to cross examine your testimony when we place you on the witness stand. He will ask you under oath in front of the jury, 'Where is the cryogenic chamber and records and what is inside that chamber?" He has video that he will show the jury, suggesting you removed the chamber."

He looked over at Robert, "I am guessing you are the second person in the video. Dobbs can't prove it is you two, but he can place that thought in the jury's heads. In all the evidence he will try and place into the record, that is the one tangible piece of evidence people will see. It doesn't prove it is you two, but Dobbs will contend that it is. Then he will turn and look at you on the stand and ask you for the truth about the video. We will object to the video being put into evidence. And whether Judge Thompson rules in our favor or not, once Dobbs tries to enter it into evidence, the jurors will have to wonder about its content. And they are smart enough to know if the prosecution is wanting to enter the video, it must have something incriminating. We are damned if the judge allows it or if he doesn't. At that moment, every eye in the courtroom will be on you. If you don't answer or say that it is not you or your son. The jury will want some proof. They will want more than just your word. And if you admit it is you two, then you will have to answer Dobbs' question; what were you doing? Either way he will have you where he wants you. Like a good chess player he will have moved you toward the position he wanted in the first place. Then it comes to the big question of creation. He will have showed the jury you have been working toward the creation of mankind and he never had to ask you straight out. It keeps him from being the prosecution who brought down the doctor with the question of creation. It keeps him free of the controversy it raises. The jury and the world will see the answers clearly through your statements or lack of. He will let people's imagination take over and not allege anything in open court. In fact, he doesn't have to ask you another question. And he knows

Downing will not approach the question either. This is a no win situation. At that point Dobbs will be ready to have the jury render a decision on the embezzlement charges. The other questions of Elizabeth's science will be for the court of public opinion. His hopes for that judgeship are still alive. He's avoided standing in front of the jury and showing if he believes man was created as it says in Genesis or in a basement lab down the road. In fact, among extreme conservatives, he probably moved up in their estimation."

Charlie paused for a moment and let all that he said sink into Elizabeth's mind. Finally he said, "That's what you risk if you go on the witness stand. You risk everything. But if you don't, they will judge you on the science question and convict you on the embezzlement charges."

Downing, who had been listening, said, "Why object to the video being entered into evidence? Why can't we simply say that we have no objection to the video and that we have reviewed it and see nothing incriminating for our witness?"

Charlie answered, "Dobbs will be all over that the minute we speak up. It's his contention it is Elizabeth."

Downing quickly responded, "I saw it and I didn't see my client anywhere in the video. Hell, as far as I know its two kids with a keg of beer that looks like a cryogenic chamber. In that fuzzy video an aluminum beer keg and a stainless steel cryogenic chamber could look the same. At least we gave the jury an alternative answer to the question raised by Dobbs. After all it is the only physical evidence other than financial facts and figures offered into evidence. At least as far as we know at this point."

Charlie looked over at Jake who responded, "The kid's got a point. If nothing else, we'll get to argue its importance in the case. Once Dobbs tries to submit it, he's committed. Then we simply say we agree on his submission, but not on the conclusion he wants to draw. At least we show the jury we don't want evidence withheld from them. We'll even portray its admission in that light. Total disclosure that proves nothing."

Charlie looked at both Elizabeth and Robert, "You realize if we go this way, he will ask you point blank, is that you in the video?"

Elizabeth looked at Charlie, "You're saying Dobbs will present this as possible evidence of my involvement in the removal of that canister?"

Charlie got a strange look on his face as he said questioningly, "Yes, but I don't think that bothers you. Is there something you want to tell me about that video? Like it's not you?"

Elizabeth's glanced at Robert and then back to Charlie, "Well, is he going to ask that?"

Charlie sighed deeply, "I guess that answers that question of who took the chamber and records."

He looked at Robert. "Dobbs has not placed your name on the witness list, so far. To be honest I don't think he realizes you were part of this. But he is no fool. When your mother testifies under oath she did not remove the canister, he's going to figure who else could have been involved. So far he has not focused much attention on the second person in the video. Your mother is the only one in his target zone. But if he asks her and she in any way moves her eyes toward you, just like she did a second ago, you are done. He doesn't have to go through your mother for information, he's got you to get to her. Young doctor, family, new career. You have a lot to lose. And he knows your mother, or at least he figures your mother will not sacrifice you to save herself."

He looked at Downing, "This changes everything. That video will trigger everything to come apart. Dobbs will figure there is something we are hiding when he submits the video. Jake, can you get into the system where they archive the lab's security videos? See if you can find anything in the video files that we can use to prevent Dobbs from using it. If there is anything, and I do mean anything, out of the ordinary, we can move to prevent Dobbs raising the issue in the courtroom in front of the jurors. The only way this works is if he knows we have something that will make his submission look bad."

Jake looked at Charlie, "What do you want me to find? That could take hours, even days to find...what am I looking for?"

Charlie responded, I know it's a long shot. He's not likely to enter it into evidence until he concludes his witnesses' testimony. We got a day or so before we have to stop him."

He looked at Robert, "You have put yourself into a bad position. What did you do with the chamber and records taken from the lab?"

Elizabeth stood up as she said, "Don't answer that. My son is not to be involved in this. You do whatever it takes to prevent him from being pulled into this."

Charlie paused before he spoke. He had obviously become upset with the situation. "Too damned late for that now! He should have thought of that before you pulled him into this situation. And I am getting tired of no answers about the chamber. You want my help, it's time to tell me where the chamber is. Our asses and credibility are on the line along with your son's career. WHERE IS THE DAMNED CHAMBER?"

From in the back, Tucker's voice broke the tension in the room. "It's gone. It is no longer in this country. It was shipped back to our long term storage facility at the university. It was included in a shipment of medical specimens used in West Africa disease research. It was a simple matter to include one more nitrogen storage chamber. I helped to make the arrangements. Elizabeth asked and I agreed."

Charlie was furious at Tucker, Elizabeth and Robert. "Jesus Christ, do you know what you have done. You stole evidence.

Even worse you shipped it out of the damned country! And worst of all you got the rest of us involved."

Downing looked at Tucker and with a questioning look, "Why have you, a man of the cloth, become involved in this? I would think this is completely against everything you believe."

Tucker stepped beside Elizabeth and placed his hand on her shoulder, 'When she contacted Angus, she told him what she had discovered. She explained what she had accomplished with the embryos and the genetic information from the Israelis. She knew that when the university's ethics board found out, they would seize her research and the records. Angus knew about the shipment of tissue samples for the West Africa Diseases research project. It had been scheduled to be shipped in a couple of weeks after arrangements were made. He simply added to the invoice a line item to include one cryogenic chamber with human embryos. He knew that if it was on the customs inventory list, there would be no problem getting it out of the country. You see, it was done all legal and proper. No one paid any attention at the time, as long as it was on the inventory documentation. This was not even a news item then. And to be honest, Angus had no idea it was going to become this big of an event. He didn't believe it would get out in the news about her research. You Yanks have the ability to make things bigger for the day's news cycles. Although he did nothing illegal or improper, he certainly could have been more open about the embryos with his own university. He's now involved Stonemore's DNA storage Ark in the argument."

Jake asked, "Why send them back to England? The government will just make your university send them back to the states. End of story and the doctor still goes to jail. The university here ends up with the research and the embryos"

Elizabeth sat up as Tucker explained, "It's not that simple. You see, in England the government some years ago recognized the science of genetic engineering and its potential for the human race. They instituted laws governing the science of DNA research on embryos. In fact they went so far as to include genetic hybrid research, bonding animal DNA with humans. England has probably done more to advance the science while at the same time trying to answer scientific and ethical questions. England has regulations to oversee this type of research and your country is… well, I guess, you are like the wild west, make your own laws. Elizabeth knew her discoveries had to be protected and the science of her research had to be held from the world. At least until the world was ready to use it responsibly. The handling of the embryos will be overseen and governed by established guidelines in England. Doctor Whitechurch is head of the government's ethic board. In Elizabeth's lab the research would have been seen as scientific property owned by the university. It would be patented and become their property. Shortly thereafter they would have a joint partnership with a bio-engineering company marketing the technology. Elizabeth's research is not hers, but the schools for which she works."

Finally Charlie asked, "So let me get this straight, your research was not working to find answers to genetic questions to help prevent diseases? You were misusing your grant money to fund your own twisted research? Is that about it?"

Elizabeth said, "I will answer all your questions, but not here. I want you to put me on the witness stand. I will answer them, but only for the record. There are things the world must know. And the only way they can hear the truth is from me. In a court of law, in front of the jury, I will answer your questions."

Downing had been trying to understand all that was being said. "Do I understand you are going to admit to everything in front of a jury? You're going to throw yourself on your sword after pleading not guilty?"

Elizabeth stood and looked him in the eyes, "I am not going to admit to any kind of a crime. That's part of what this is all about. Is what I have done right or wrong? If the wrong people gain access to the secrets of my research, they will use it to benefit only those who can afford it. But used with no rules and regulations as to how far it can be taken. My time on the witness stand will make people ask questions, not so much about my discoveries, but about the ethics and legality of my science, much like England did twenty years ago. Don't you see? My work can…"

She stopped herself abruptly, "NO! I want to testify to what my life's work has accomplished in front of the courts."

Everyone looked at Charlie, waiting for a response to her statement. "If that's what you want and you are willing to accept the consequences of your actions, I cannot stop you. But, we are still going to have to let Dobbs present his case against you. If I understand correctly, you are not wanting to change your plea to guilty and confess, you want to explain yourself and your work. And that explanation will back you're not guilty plea? Have I got it right?"

Elizabeth quickly responded, "Yes, that's about right. I stole no funds from the grants or endowments as the government alleges. But I do realize now my research deviated from the directions the grants were written for, but not to defraud the program, just to find answers that needed to be found. As to the embryos, I hope my explanation will be understood by the jury."

Tucker intervened as she spoke, "Can your explanation answer the questions of the cryogenic chamber and it's being sent back to England?"

She turned and looked at Tucker, "Do you think Angus can make the trip? He can be…well, a character witness for me. He can also explain your country's laws concerning this type of research. He has been part of the Human Fertilization and Embryology Authority since it first began. He can at least shed light on how your country has made great progress in regulating this type of research. He also may be able to explain my reluctance to allow my research to be used in this country."

Downing said quickly, "Because England has laws and America doesn't, his testimony may have no bearing on your case. In fact his answers may make things worse."

Charlie said, "If you are trying to force our government to adopt those rules, you are way in over your head. They will not pay any attention to your answers in the courtroom. As far as the government is concerned, you are an embezzler. End of story."

Elizabeth responded. "I know they may not heed my warnings, but when the news media gets the whole story, they will ask with a voice much greater than mine. There will have to be answers. Until then, my research documentation and the frozen embryos will not see the light of day."

Charlie looked straight at her as he said, "Even if you go to jail?"

Elizabeth stared back, "I am ready to make the sacrifice to tell the world of the discoveries of my science and the benefits it could create for all."

Charlie sat down and looked at her, "Believe it or not, I kind of wished you had been creating a hairy elephant in your labs."

Everyone smiled at his comments.

Chapter 24
The Trial Begins

Charlie was awakened by someone knocking on his door. In his darkened room he squinted to see the time on the clock, it showed four thirty. As his eyes cleared the knocking continued. Finally he said to whoever had awaken him, "Just a minute, let me get some clothes on."

He managed to find his pants and make his way to the door. As he approached he could hear Jake saying, "Charlie, let me in I've got some information for you."

Finally Charlie opened the door and, rubbing his eyes said, "What the hell is so important you had to wake me?"

Jake said in response. "Wake you, hell I have not even been to bed. It took longer to get approval to look at the footage than it did to find what we were looking for. You are going to love this."

Charlie sat down in the chair and forced himself to awaken and talk with Jake. "Okay, what did you find?"

"You remember the video of those two people stealing the cryogenic canister? Well, that tunnel is the main late night highway for a lot of weird activities. The college kids know they can get into the lab building without getting caught most of the time. The lower levels of the lab are unsecured except for the cameras. And if they are detected, by the time security gets there, they are gone. I have video of them stealing dry ice, rolls of toilet paper and empty nitrogen canisters. The dry ice was used for a Halloween party. The toilet paper was used to decorate the trees in front of the administration building after some rule changes on drinking in the dorms was proposed. And the nitrogen canisters were apparently cut in half and made into beer coolers. The videos and the follow up reports are all contained in the security section of the lab's footage. If Dobbs tries to use his video, we will be able to show that security in the basement of the lab building has some major problems. There is no way Dobbs would allow his video into evidence with what we have to counter his video. Especially since it doesn't even show who the people are in the tunnel. Oh, one more thing, I saw Tucker a short time ago. He has spoken with Whitechurch and he has agreed to testify on behalf of Elizabeth. Or at least he could explain some of the work being done in bio-genetics. He figures it could be of some help for the jury to understand where England is in its research and how they have established guidelines for the science. He should be here by late this evening. It looks like things are coming together."

Charlie sat back on the bed while letting the nighttime cobwebs clear. "I am glad you feel that way. But this case has taken so many turns, I'm not sure how to handle it. I still believe letting Dobbs make his presentation with each of his witnesses helps to give us some basics for Elizabeth's work. In cross examination Downing can clarify any damaging information concerning Elizabeth's directing of grant monies. Still, the real problem will be Elizabeth's testimony. Putting her on the witness stand gives Dobbs a shot at questioning her. She can be confrontational and angry when pressed for answers, she does not want to answer. If the jury sees that side of her it will not be in her favor to gain their trust. And besides that, I still don't believe she has told us the entire truth. If Dobbs decides to ask about the substance of her research into DNA and she doesn't get it right, she could end up looking like a science doctor gone mad with her work. Some of the media stories are already going down that path. You know the jury has been instructed not to read or watch news reports…most will do as the judge told them…but we both know at least one will watch the news. When they go into deliberation, they could say something to influence the other jurors."

Charlie looked at the clock on the night stand. By now it was reading a little before five. "I am going to get a shower and go over to the coffee shop across the street and get some breakfast. We should be there when Judge Thompson brings those protesters in this morning and see how he handles their hearings. After all, a couple hours later we will be in front of him…I would like to know what his mood is after dealing with those people."

With that Jake excused himself and told Charlie he would meet him for breakfast. Charlie told him to contact the others and see if they wanted to join them.

An hour later he was sitting in the coffee shop as the others arrived. He looked at Downing, "Are you coming with me to court to hear how the judge deals with those protestors from yesterday?"

Downing replied, "Absolutely, I want to see his mood. Especially if they get a little mouthy with him. I am thinking at least one of them will challenge his authority."

Jake quickly responded, "I've seen him before when he believes someone has shown his court contempt. He has zero tolerance for anyone who challenges his authority. They have only one chance to apologize. If they don't, they won't like the alternative."

After that exchange of conversations, the group ordered breakfast and talked about what they expected that day in court. They discussed their expectations that Dobbs would be introducing information about grant request programs and the way the university applied the funds. The jurors will be unlikely to have much knowledge of how grants are requested and administered. Dobbs will give them enough information to at least be comfortable following testimony of his expert witnesses. That way they will understand what the government is alleging and how she abused her authority inside the programs. He will try and present a very narrow framework for the grant's application and how they are to be used. The government will try and box Elizabeth's authorized research into a defined course of study. Charlie knew Downing could handle any damaging testimony with rebuttal questions and give the jurors their side of the situation. They did not expect any surprises in court today.

After breakfast they returned to their rooms to pick up their briefcases. Everyone except Elizabeth met in the lobby at seven thirty. Charlie had told Elizabeth he did not want her there until a half hour before the trial would begin. The marshals would come to her room and escort her to the courthouse by the lower security garage entrance. A short time later they left for the federal building.

As Charlie and the others entered the courthouse, they were met with angry shouts from those who were there to disrupt the proceedings. News reporters were pushing microphones into their faces while asking questions. As they pushed their way to the second floor Jake said, "If you don't mind, tomorrow I'll ride in with Elizabeth and the marshals."

After a few moments, building security formed a barrier around the group as they made their way into the courtroom. Just inside the doors Charlie used his briefcase as a pointer to indicate the group should sit near the back. Along the front row on the left side were the dozen protestors who had been taken into custody yesterday in court. Standing in front of them was Reverend Jedidiah Cain. He was leading them in prayer. Charlie and the others were too far away to hear exactly what he was saying.

It was close to eight fifteen when the bailiff announced to the near empty courtroom, "All rise, the Honorable Judge Harold Thompson presiding."

The judge entered and sat down at his bench, looking at the protestors. "All right, yesterday I found each of you in contempt of this court. You disrupted the proceedings and chose not to follow my orders. Consequently each of you spent the night in jail. I would hope the time spent in the care of the court system has taught you to understand the error of your ways. I want it understood by each of you that I understand your right to protest, but not in my courtroom during a trial."

He paused and looked at each defendant before he spoke, "Before I pronounce my decision as to what I intend to do with each of you, is there anything you would like to say to the court. I will listen to each of you one at a time."

Charlie leaned over to Downing and whispered, "He's going to give them a chance to apologize to the court. If they are smart they will take his offer."

No one moved for a few seconds, then Reverend Cain stood and asked the judge, "Your honor, may I speak for those people you have been detained for their beliefs?"

Judge Thompson sat back in his chair, clearly not wanting to grant his request for a platform to further his beliefs. As he thought to himself, he remembered that so far Cain had been respectful and polite when addressing the court.

Finally after a short pause, Judge Thompson replied, "Reverend Cain is it? I will at this time allow you to answer for these people. That is if they are willing to accept my judgment after you speak for them. I would remind you before you begin, this hearing is about their contempt for this court. Your words will help to define any further punishment that I feel is necessary."

After hearing the judge's statement, the group of people mumbled their responses as they shook their heads they agreed, Cain could speak for them.

Cain's chest swelled up as he stepped before the courtroom. He turned to the group and gave them a smile before turning to the judge. "Your honor, these people were not protesting when they entered your court yesterday. The time of protest has passed. They felt it was time for actions. In your courtroom was a woman who calls herself a doctor of medicine and yet she has brought such an abomination onto this earth, the only course of action is to point out her crimes before God. The courts of man cannot judge a crime so horrendous it is unspeakable among God fearing men and women. These people...."

Judge Thompson smacked his gavel onto his desk and said loudly, "ENOUGH! I am here to accept an apology from each and every one of the people who sit behind you. I will not let them, or you lecture this court on its obligations within the laws of this country. Be clear, each of them can stand and apologize or I will find them one thousand dollars apiece. And if they again show contempt for the court, I will lock them up and forget where they are for the next six months. Do you understand me Reverend Cain?!!!"

Cain stood as if he were high on a great pulpit, "My followers will not apologize. Our church will pay each of their fines. I will also have the attorneys for the church file an injunction against this court for religious bias."

Thompson sat staring at Cain. In a moment he said, "It is your right to file whatever legal papers you may choose to file. It is of little consequence to me. And you can pay their fines if you wish to. Let's be very clear here, if there are any more protests that disrupt my courtroom and I believe... no, that's not correct, even if I suspect you had anything to do with it, I will summon you before this court. And I suggest you get a real attorney to represent you. Not God...nor yourself... a real attorney."

Thompson leaned forward and shook his finger at Cain, "If you come before me again, you will get your chance at the television cameras, as they load you in the police vehicle and take you away. Your veiled threats supposedly sanctioned by God hold no credence in this court."

Cain started to say something else when Thompson struck his gavel and said, "Pay the clerk and clear those people out of my courtroom." With that he disappeared into his back office.

When the judge was clear of the courtroom, Cain turned and looked at Charlie, "Mr. Morgan, you never did answer the doctor's question, do you believe in GOD!"

Charlie sat staring at Cain and gave him no response. Each of the protestors stared at Charlie and the others as if they were the Devils representatives on earth.

As they stood to leave Cain shook his finger once more at Charlie. "You better believe, I'll warn you, you better start to believe. Because, I don't even think almighty God himself can save you from eternal damnation if you continue to defend that BABYLONIAN WHORE!"

Over the next hour, Charlie and Downing reviewed their case. About nine thirty Dobbs and his team entered the courtroom. Dobbs gave Charlie a nod as a way of saying good morning. Shortly thereafter an assortment of religious men entered the courtroom and sat near the back. They made no small talk among themselves. Each of the stern faced men sat quietly and looked forward toward the front of the courtroom.

Over the next few minutes the courtroom filled with reporters and those who were lucky enough to be chosen for the main courtroom seating. Those who were not so lucky were relegated to sit in the courtroom just down the hall way. They were to view the proceedings on closed circuit television. The bailiffs in the auxiliary courtroom made sure they understood no electronic recording of the closed circuit television images would be allowed. The same restrictions were in place in the main courtroom.

At a quarter til ten, the two marshals escorted Elizabeth through the side door.

Over the next fifteen minutes the courtroom settled itself as people waited for the day's proceeding to begin. A few minutes before the hour the jury was escorted into the room.

Precisely at ten o'clock Judge Thompson entered the back of the courtroom, his bailiff announced, "All rise, the Honorable Judge Harold Thompson's court is in order."

As Thompson sat, he motioned for the room to sit down. He shuffled through a few papers before looking out at his court. "I believe we are ready to pick up where we left off yesterday. Is the prosecution prepared to begin with its first witness?"

Lisa Williams, assistant prosecutor working with Dobbs, was to handle the day's questioning. The government's first witness would explain federal grant programs for the members of the jury. She replied to Thompson's question, "Yes, your honor, we are prepared to call our first witness."

He motioned toward her with his hand, indicating that she should get about her work.

Ms. Williams declared to the court, "The government calls to the stand, Mr. Patrick Hassam."

A tall thin man in his middle fifties entered the courtroom from the side door. He made his way toward the witness stand and stood waiting to take the oath. The court clerk administered the oath and Hassam sat down.

Ms. Williams stood and proceeded to walked in front of the prosecution's table. "Mr. Hassam, would you please state your full name for the court?"

"Patrick Hamilton Hassam."

"And for the record, Mr. Hassam, could you explain your background and education to the court?'

"Yes, I am a graduate of the University of Chicago. I have a Masters in Business from Harvard. I did my graduate work in Washington for the Commerce Department working in the grant research programs. I eventually left government work and began writing grant applications for colleges, primarily those colleges and universities looking to received monies for medical research."

Williams asked, "Do you work with applications from different schools and are you primarily a grant writer for those with medical research programs?"

"Yes, my primary grant writing is done for colleges and universities that are considered leaders in research into cancers. In order for me to make applications for grants, it requires me to keep up to date on new grant programs that are being proposed in government and even sometimes private monies. I first find out how much money is going to be available. Next, if the money directed towards a special cancer or disease. Then I check and see if the university has received money before under similar grant applications. Each university is rated by their success in research. Areas to consider before grants are written…does the school have a good rating based on past success? Does the program have some of the top graduates applying to work in their research programs?"

Williams interrupted him and asked, "Do past graduate applications have influence on the grant request?"

"Absolutely. Research programs that draw the brightest and best of their particular fields of studies are very important. If the project brings top graduates, that helps to bring in top money. You have to remember that often federal grant money is coupled with private money. Non-government businesses want to be sure they get a return on their investments. A program that is recognized as a leader in the community of scientific research helps to guarantee results. All these things must be kept up to date before applications can even begin to be applied for."

"In basic terms, can you describe to the jury how a grant is requested?"

"When I learn that a new source of funding is going to be made available for grant programs… lets' just say in the president's state of the union address he speaks to a new effort to find answers to a specific cancer such as breast cancer. I now know that monies have been set aside to be directed toward research. I will quickly find out how much money and the guidelines to apply. I then contact the different universities and colleges and see if they are interested, based on the grant requirements."

Ms. Williams thought for a moment before asking the next question. "Well, Mr. Hassam, if you contact all the schools you are working with, how do you determine how much each school should apply for?"

"Sometimes the money is already blocked out into amounts. Other times it is for me making the application to apply for a special amount."

"Then I would assume when you apply for a specific amount you have some internal criteria in your applications for funding?"

"Yes, past performances and top ratings makes a difference. When I first got into the business of writing grant programs, each request was evaluated on the application itself. But today grants are evaluated more by past performance of those applying for monies. Doctor Mathenson's program carries one of the top ratings in the field of DNA research."

"So to be clear, you would tell your applicants that past performance is the gateway to gaining future grant monies?"

"Yes and no. It is hard to quantify a grant purely on success of the past, although that is a big factor in looking at the projects. In science and medicine, results are not always tied to a

"Eureka" moment. Some success in one area of the research may lead to small steps forward over the years. In the past ten years medical science has made great strides in many different cancers. Yet people are still dying every day. For them, the science had not come soon enough. But did one of the research programs that received money make a small discovery that opened the door for a new treatment? Yes, that is possible. I have to evaluate all these factors when making the applications for money for each university."

Ms. Williams turned towards Elizabeth, "Mr. Hassam are you familiar with the defendant in this case, Doctor Elizabeth Mathenson?"

Mr. Hassam gave a quick glance toward the doctor and Charlie could see a hint of a smile on his face as he said, "Yes, I know Doctor Mathenson."

Ms. Williams looked at the papers she held in her hand before asking the next question. "Mr. Hassam, I see from the records that I entered into evidence, your name along with Doctor Mathenson appears on all the grant applications going back fifteen years. Is that correct?"

Hassam thought for a moment before answering, 'Yes, I would guess that is about right. Without reviewing my own records I would have to say that is by my recollections about right."

"Okay Mr. Hassam, we'll say you have been working together for many years in making grant applications with Doctor Mathenson. So you both work to apply for money for the program that she oversees at her research lab?"

Hassam adjusted himself in the chair as he paused before answering. "My job is to keep up with grants when they become available and to write the grant application for my clients. I have known Doctor Mathenson for many years and our interaction is mostly… well I would have to say limited to letting her know, along with her advisory board, that monies are going to be available. The entire board reviews the guidelines of the grant and decides if I should go ahead. I get paid for writing the applications. Not by how much the program gets. The incentive for me is to write a grant application without any errors. If I do my job correctly, the programs receives the monies. That's pretty much how it works. I don't oversee how the money is spent inside the program or how members of the project directed its funding."

Williams paused before asking, "So to clarify, your testimony to the jury is…your responsibility is complete when the grant is applied for?"

"Yes. I assume that if the program receives the money, they will follow the guidelines of the grant application. I don't review the use of the money myself. Each program is set up to review its own course of action. They file reports as to goals applied for under the grant and if those goals were met. This is done with a survey report where the areas of questions are rated one to ten. The final number determines the rating of the program."

Williams quickly followed up with, "In the case of Doctor Mathenson, the money and how it was spent is outside your normal guidelines. You're telling the jury, use of the grant money is none of your professional concern?"

Hassam bristled at the question's implications, "That's not what I said. You are twisting my words. Each program has a review board internal to itself. And it is not unusual for the government to do audits of the programs and monies and how they are being spent. These reviews contribute to the rating of the program. Any question about use of money would not only cause a review, it would drop a program's rating. A loss of rating not only makes money difficult to obtain, it is a red flag for post-graduates making applications to work in the program. I am concerned not only with any problems with Dr.

Mathenson's programs, but with any of my clients losing a high rating. As I said, I get paid by writing the applications. She loses her credibility and it is unlikely that I would be asked to write an application for her research program. I deal with well-established universities and colleges. They are not about cheating on grant projects. Credibility for them and myself is keystone to the process."

Williams quickly asked, "Do you have any knowledge whether Doctor Mathenson's programs were under scrutiny by the government as to how the monies were being spent?"

"No, I have no knowledge that she or her program was under any scrutiny for any reason. The university has always adhered to the highest standards."

"One more thing Mr. Hassam, you said you know Doctor Mathenson. What sort of person do you believe her to be?"

Downing stood quickly and said, "Your Honor, I object to the prosecution's question. It asked for his opinion. That opinion could be driven by many different factors that may have little or nothing to do with the case before this court. My colleague is trying to box Mr. Hassam into giving an opinion to the jury, not based on facts in the case, but on personal feelings."

Judge Thompson sat for a moment before answering, "I am going to allow the witness to answer. He has established that he has known the doctor for several years. It may only be his opinion, but opinions regarding her actions toward the grant programs can be useful for the jury to try and understand her approach to the directing of the monies. Your objection is noted, but overruled."

Downing looked at Ms. Williams and gave her a bit of a grin as he sat down. She paused for a minute collecting her thoughts about the judge's response. She was more puzzled as to why Downing grinned at her. Finally she said to Mr. Hassam, "Once again, what type of person do you believe Doctor Mathenson to be?"

Hassam looked at the jury and then at Ms. Williams before answering, "Doctor Mathenson is likely the most driven person I have known in my life. For the most part she pays very little attention to the monies and the political correctness of working with members of the board inside her own program. I have tried to tell her over the years that if she could be a little nicer to the governing board, they would be easier to work with. She is driven to find answers to cancer. She believes the answers are within each of our DNA sequences. She is so driven, it's as if she had some personal agenda. But, as far as I know, she is only dedicated to finding answers to life's greatest medical mysteries purely for the answers. But when it comes to her social skills, she is difficult to work with."

"Mr. Hassam, do you think she is capable of being dishonest when it comes to money in the programs?"

Before Downing could object, Hassam said, "Absolutely not."

For the next two hours Ms. Williams questioned Hassam on the different aspects of grant projects and money, mostly to make sure the jury had a full understanding of how the projects worked and how the university administered them for Mathenson's research. Finally she looked at Dobbs who tipped his head, indicating he was satisfied with her presentation. She had clearly given the jury an understanding of the information she had presented."

At that point she turned to Judge Thompson. "Your honor, I have no further questions at this time. We would, however, ask that Mr. Hassam not be dismissed from the proceeding, in the event his expertise would be needed later in the trial."

Judge Thompson nodded in agreement before looking toward the defense table. "Mr. Downing, do you need a few minutes to prepare your response to Mr. Hassam's statements?"

Downing stood and said, "No your honor, we are prepared to redirect while the jury's memories are still fresh from the witness's testimony."

Judge Thompson nodded his head and said, "Then you may proceed."

Downing walked to the center of the room. He looked at the jury and smiled as he said, "That sure was a lot of information. I want to make sure you remember all that Mr. Hassam said about grants and how they work. I know he sure filled my head with a lot of financial information. Looking at your faces during his presentation, I suspect he filled yours also with a lot of accounting ideas and terminology."

With Downing's short statement, Dobbs stood and looked at the judge with an odd questioning look on his face, "Your honor, must we endure Mr. Downing's second opening statement. He has made his opening, can we now get along with the questioning of the witness?"

Thompson looked over his glasses down at the defense counsel, "Mr. Downing, can we please proceed with your questions?"

Downing again looked at the jury and grinned as he said, "Yes, your honor."

He turned his attention to Mr. Hassam and paused with a smile before asking his next question. "Do you think Doctor Mathenson is a crook?"

Dobbs abruptly stood and blurted out, "I object! I wholeheartedly object to such a statement about his own client. Why I have never come up against a defense counsel who makes a suggesting accusation about his own client…and in front of the jury!"

Thompson sat back and breathed deeply as he looked over at Charlie who had not moved a muscle in the last five minutes. His expression gave no indication if he had a hand in Downing's opening question. He then turned to Downing, "Mr. Downing, you do realize the doctor is your client?"

"Yes, your honor. But my question gets right to the heart of their witnesses three hour response to the prosecution's questions. Ms. Williams strategically maneuvered her answers toward the direction that she used the money illegally. My question is meant to be direct and blunt. I want the jury to understand what Mr. Hassam's answer is when it comes to my client's guilt. You did give the prosecution latitude in allowing Mr. Hassam to state his opinion about my client. I believe him to be an honest man. He must have an opinion as to why he is here and if she did misuse the funds, even if it was not his responsibility to police their use. After all, the prosecution has more than established that he is the paper man in the funds she is alleged to have embezzled."

Thompson sat looking at Downing and then at the jury. He then glanced at Dobbs as he said. "Mr. Hassam you may answer the question."

Hassam hesitated before he said, "Could you please repeat the question?"

Downing looked him square in the eyes as he said in a questioning tone, "Do you believe Doctor Mathenson is a crook?"

Without hesitation Hassam said, "No, I do not."

There was an audible murmur in the courtroom as he finished his answer.

Downing turned and looked at the jury as he said, "Can you explain that answer to the jury? An answer of no, is definite when it comes to the question before the court."

"I am a grant writer, not an accountant. The questions I have been asked are about procedures and where the grant money was spent. Definitions of grant requests and how broad those definitions can be interpreted. Not once have I been asked how money was hidden or lost from the projects at the university. As far as I can tell from the questions, they know where the money is, they just don't like how it was spent. Doctor Mathenson and her team are not embezzlers. They don't necessarily worry about every penny and proper protocol for grant applications. But I suspect this is not about their illegal actions, it is about goals they set for the programs. And as far as I can tell, this would not have even been a question before this court if she had played ball with the politicians."

Dobbs stood and vigorously objected, "Your honor, Mr. Downing is leading the witness to ramble on about things that he has no knowledge of. He is attempting to enter hearsay through the witness's testimony. We ask that his statements be stricken from the record and the jury be admonished to disregard his statements."

Thompson looked at Downing, "This is a federal court. I expect you to be a little more... how do I say this...a little more professional. I will not tolerate any more theatrics, is that understood?"

"Yes your honor."

Thompson looked at the jury who was staring at him waiting for his response and wondering if he was going to tell them to forget what Hassan had just said. He did not. Instead he turned to Downing, "Do you have any more reasonable questions for Mr. Hassam at this time?"

"No your honor, with the same understanding that the prosecution has stipulated, that we may possibly need to recall this witness. I would make the same request, in the event any new evidence would come to light and I feel Mr. Hassam could clarify it for the jury."

Thompson looked up at the clock on the rear wall. "It is one-thirty, we will reconvene at three. Court is recessed for now." With that he stood and left the courtroom.

Jake stepped up and placed his hand on Downing's shoulder as he said, "Jesus James! 'Is she a crook?' That's a pretty direct question. Did they teach you that in law school?"

Downing half grinned, "Nah, my grandfather taught me that. He used to say, "Don't pussy foot around, just ask the damn question. So the most direct way in this case was to ask the damned question. I figure he was the first witness the jury heard from. The prosecution was putting them to sleep with her questions. All the time, she kept moving his answers closer to blame for Elizabeth. She was implying with her questions that Elizabeth knew something about the money. Between the boredom of grant protocol questions and her implied implications, the jury was falling asleep believing her fairy tale. All I did was to ask the damn question and wake the jury, is she a crook? He said no. That's what they will remember about that three-hour session. The person closest to her and the money said she was not a crook. Everything else will be forgotten. I gambled I could use their expert witness to be a character witness for us. I hope it worked."

Elizabeth looked up at Charlie, "What's next?"

Charlie closed his briefcase. "Well according to their witness list, pretty much the same. They will have the controllers from the university go over the money. They have two members of the governing board to testify. But it all comes down to the evidence Hassam put into place with his testimony. The money is there, it is just a question of how it was directed inside the project. And in fact each of them have far more control over how the project monies was used. Hassam stated several times he had no part in the money's use once the grant had been approved. He is just the grant writer. So far they haven't any real evidence showing where money was taken, or used illegally, they don't have much of a case when it comes to the money."

Charlie took Downing aside as he said, "You got away with your direct question, but you were pretty sure Hassam would say she was not a crook. What if he indicated he wasn't sure or even hesitated with his answer? Your direct question could have only been answered one way, with a direct NO! Anything else and our case would have been over. Leave your grandfather's advice out of the courtroom from now on, is that understood? Unless you know what the witness is likely to say for sure, don't take the risk. Is that clear? It's not your ass on the line, it is Elizabeth's."

As Downing shook his head that he understood Charlie's comment, the judge's clerk handed him an envelope. He opened it and read it quietly, he then handed it to Charlie. Elizabeth asked, "What does it say?"

Charlie handed it to her. There was one name on the list, Clay Gordon. He looked up at Downing as he said, "We did give the prosecution our potential new witness which included Whitechurch didn't we? If they intend to put Gordon on the stand, they intend to open the evidence up to more than just the money. I am sure of it. Gordon would not reveal his Israeli secret for just questions of grant money. He's going to open the dialog up to the question of the Israeli DNA. We'll need Whitechurch to be our expert witness."

Downing indicated that they had given the court his name as a potential defense witness.

Jake looked at the paper before he said, "Is Gordon in response to our requesting Whitechurch as an additional witness or is he going to testify on what he knows about the case? He will not be testifying about the money… will he?"

Charlie looked at Downing, "No, I don't believe so. I'm betting Clay is behind this embezzlement case against Elizabeth. Remember, she told us that last summer she talked with Clay about how to prevent the university from gaining patent rights for her work. She had known Clay for years and they were good friends. She had no reason to suspect he would betray her questions. When she spoke with Clay, she must have said something that opened a question for him about missing DNA. After their conversation, I'm betting Clay contacted his government as a matter of routine. When he mentioned the samples from their country, no red flags were yet raised. Then someone in the security ministry must have discovered Elizabeth was in Israel on a science visa. They looked a little further into the question of her whereabouts and they discovered the samples were missing. That's when someone in their country put two and two together, and they realized her research involved DNA from their labs. I'm betting internally they were not yet aware the lab's curator sold off the duplicates. He just intended to clear out what he believed were unneeded extra samples. I'm betting at that point something very important was discovered about the batch of missing samples. So important, the Israeli government got involved and contacted the state department. They also instructed Clay to find them and try and gain their return, even if he had to tip his hand about his own political situation. He must have been instructed to go so far as to testify on the witness stand. Only thing I don't know is how much he's willing to reveal on the stand. He has to know we will cross examine him and we know he is an Israeli agent. His testifying will expose his country and himself to whatever his true mission has been. If he's willing to give that all up and risk being arrested, and even deported, he's going after those DNA samples Elizabeth took out of Israel at any cost. He doesn't care about the money or her. But he knows once she admits to having the Israeli DNA samples, they can petition our government for their return, even if it cost him everything."

Tucker, who had sat quietly, finally commented. "There is no doubt that is what he's after, but he does not yet know they are gone from her lab. He is giving up everything to make a play for those samples. I'm betting his government is backing his taking part in the trial. And I am guessing your government has some idea why he's agreed to be part of the effort to get them back. Don't forget about that off the record meeting when we returned from England. We were met by your country's Attorney General. I am guessing that never happens, but in this case it did. The Israelis want them badly. And your government wants very much for them to have them back and the problem solved. Your government has the upper hand in the matter, something they could use for leverage in some future power play."

Jake blurted out, "Yea, they want them, but just wait until he finds out they are already gone. It's going to hit the fan. And in all likelihood Elizabeth is going to get hit with the flying debris. Like it or not, Clay and his government are willing to expose themselves just to regain those samples. What are they going to do when they find out they are gone?"

Downing said, "If they don't know, then I am guessing that Dobbs has not told them he suspects that canister was taken from the lab. He also has to know the reason Clay is willing to break his cover and testify in open court. If I didn't know better, I am beginning to think our government wants the Israelis to reveal their part in this. The justice department along with the Attorney General, knows what cards are not showing to the Israelis. Maybe they are after something bigger than we know. Everyone has something they are not telling and Elizabeth is caught in the middle with her work."

Charlie picked up the file of information that Jake had given him along with the security footage. He acted as if he was weighing the value of their contents. "Maybe, just maybe we can force Dobbs' hand by trying to submit this information into evidence at the same time he offers his video evidence. Thompson will reject our efforts and remind Downing he can submit whatever we want when it comes to the defense's turn. It's a major breach of legal ethics, but it will hopefully set things into motion."

Charlie turned to Downing, "The judge is going to give you one hell of an ass chewing in front of the courtroom. Hopefully he'll mistakenly believe you submitting new evidence, along with your brash methods, made you forget about proper evidence presentation. But Dobbs…he'll know, or at least suspect, we have something to discredit his video. We have a chance to throw him off his game and maybe even get him to tip his hand. We need to know what game he is playing. Is he working with the Israelis or is he being forced to be part of some political game from the Justice Department? This move just might do it. But, it may cost us the wrath of Thompson. If he figures out we played him to prevent the prosecution from submitting evidence, he'll blow a gasket. May even declare mistrial. If Dobbs doesn't go crazy at the same time, he's being forced to play the Justice Department's game. We'll learn a lot, but risk everything in the case so far."

Elizabeth interrupted the others' conversation, "Excuse me, but won't Clay change what he is going to testify about if he finds out about the missing samples. If I understand correctly, tomorrow morning the prosecution is going to have the head of security testify and introduce the video showing people stealing the canister. Won't Clay shy away from revealing himself for samples not in his reach?"

Charlie paused as he responded to Elizabeth's question. "Well doctor, are you ready to tell us about the video?"

Elizabeth stiffened her lip as she said, "I was just asking a question about Clay. He's the one who has a vested interest in the samples and the evidence of the canister. I would think you had better be prepared if he is not willing to acknowledge Israel's connection in this case."

Charlie said, "We'll see what tomorrow brings. Remember, I am only speculating about Clay's involvement at this point."

After Charlie's comments, the others continue discussing their strategy for the case both with Clay offering to reveal his country's part along with his own or denying they had any involvement at all. Only his testimony would reveal the answers. Before the group realized, time had slipped away and it was time to reconvene the afternoon court session.

Charlie picked up his briefcase, "Its near three; we better get back into court and make our play with the video evidence. Dobbs indicated that he was going to call the university's controllers to testify. I have looked over their depositions and I see no real problems with what they have said so far. We will have to respond in cross with explanation about Elizabeth's use of money outside the program, her spending money in Korea and Africa. There is no documentation that her work in those countries had any effect on the research in her labs. We'll hear what the controllers have to say and ask for a recess to collect our questions."

Charlie turned toward Elizabeth, "We're going to need you to listen to the testimony of the controllers. It is important we answer any questions that leave your trips and expenditures in question with the jury. It would be good if you keep a few notes to help us with our cross examination of the controllers."

Downing shuffled through the papers in front of him until he found the controllers' names. He looked up at Elizabeth, "The first witness will be Carla Hinson. She is a twenty-year university employee with the controller's office. Have you had much experience with her and are you on good relations with her?"

Elizabeth paused before she answered. "She's a pain in my ass. Questions everything. Most of the time her questions are about my poor documentation of the finances surrounding my project. She was particularly frustrated by my trips to Korea. Expect her to be detailed in her explanations."

Downing quickly asked, "Do you get along with her?"

Elizabeth grinned as she thought about the question. "We get along. I don't think of her as someone who's out to get me. She is more about keeping her documentation correctly filled out. She is one of those people who arranges everything on her desk in perfect order. Pencils are sharpened to the same length and arranged evenly on her desk. A neat freak. But wouldn't think she has an axe to grind with me, at least personally."

Charlie directed the group to return to the courtroom. As they walked in the clock showed three. The judge's bailiff opened the door to the judge's chambers and nodded his head that people were now in place.

The judge must have been standing just beyond the door as he entered the courtroom quickly. As he sat down he looked at the defense table and looked at his watch. Charlie understood the judge doesn't like to be kept waiting, even for ten seconds.

Thompson put on his glasses and sat back in his chair as he said, "Mr. Dobbs, are you ready to proceed?"

"He stood and said, "Yes you're Honor. The prosecution calls Carla Hinson."

Carla Hinson was dressed in a black business suit and wore her hair on top of her head. Her dark rimmed glasses gave the appearance they were from the sixties. She carried herself straight and proper. Her appearance seemed stiff as she stood and took the oath.

Dobbs moved toward the witness stand as he asked her to describe the position she was responsible for in the university's controller's office. Hinson described her job as if she was reading from an instructional manual. She made sure to dot every "I" and cross every "T" as she explained her job function.

Finally, after nearly an hour and a half testimony about her credentials and job responsibilities, Dobbs got into the questions concerning Elizabeth's project and its finances.

"Ms. Hinson, could you explain to the jury any concerns you raised with Dr. Mathenson concerning her expenditures of the grant funding of her project?"

Hinson turned and looked directly at Elizabeth before answering the question. "Doctor Mathenson is like so many others at the university. They believe they have a check book with no limitations. They never care about the balance of funds, in or out."

Dobbs then asked, "Would you consider her to be problematic with her use of money?"

Hinson hesitated for a moment, "Yes, unlike many of the other programs I oversee, her expenditures are scattered around the world and often hard to keep track of. She often fails to submit proper paperwork or even advise our office of the expenses she creates."

Dobbs went on for nearly forty minutes of questioning Hinson about different questionable expenditures. Finally Dobbs raised the question of money allocated to a research lab in Israel.

"Ms. Hinson, I see from your deposition that you were particularly concerned about a large payment, I believe it was thirty-five thousand dollars, to a lab in Israel. Is that correct?"

"Yes. For one thing I had no advance notice the Doctor Mathenson was even traveling to Israel for any kind of business. Then my office gets a purchase order for payment from the lab for that large sum of money. Expenditures of that size are supposed to be brought up for review prior to being allowed. When I tried to get some verification of the funds and their use, Mathenson told me to just pay it. She said she would give me a complete accounting and documentation at a later date. As I expected, she was less than forthcoming in providing the required information. In fact it came up three times in department board meetings. She failed to attend or even send documentation."

Dobbs looked at his list of questions. "All right Ms. Hinson, let's move on to another question that you had about her expenditures. I believe they were for some cryogenic equipment." He handed her an invoice. "Your honor we would enter this document, label exhibit "M", into evidence for the prosecution. Ms. Hinson, could you tell the court what this invoice is for?"

Hinson took the paper and adjusted her glasses as she briefly looked at the document. Then she handed it back to Dobbs as she said, "Yes, this is an invoice from a medical salvage company in Alabama. It shows where Doctor Mathenson purchased one embryo-cryogenic storage canister. We received no health certificates or any other documentation concerning this storage chamber. I had no idea if it was of a biohazardous nature. Doctor Mathenson's lab already has an excess inventory of one hundred bio-storage chambers. Inventory that we must account for each audit. I saw no reason without explanation why we needed to add to this inventory. The expenditure of funds for this, without proper documentation was not needed in my opinion. Again when I confronted Doctor Mathenson about the purchase, she brushed me off and mumbled that she would complete the proper documentation."

"Did she?"

"No, she never did."

Dobbs walked close to the jury box before asking his next question. "Do you know if the cryo-chamber was empty?"

Hinson quickly responded, "My office received no health certificates, canister serial numbers or manifest for any contents that may have still been inside the chamber. As I said, Doctor

Mathenson gave us nothing, except the purchase order to pay the salvage company. Oh yes, we also received a few months later a bill of lading for transporting the chamber from the medical storage facility to the university. The invoice did state that the chamber was attached to a liquid nitrogen supply container. It also gave instructions if the gauge indicated low supply, contact a number listed on the tag."

Dobbs, "Did you recognize the phone number?"

Hinson, "Yes, it was Doctor Mathenson's office number. I have called it enough to know that it is hers."

Dobbs, "And the alert to call the number in the event of low pressure, what does that mean to you?"

Hinson, "I am not a scientist or familiar with shipping proceeds, but in my laymen's understanding, the canister still contained something that needed to be kept frozen. Part of the paperwork was a temperature report from the shipping company. Every four hours the canister's temperature was to be checked and recorded for the records. I have no professional understanding of why that was necessary, but I am guessing to make sure the container's contents were kept at proper temperature during shipment. After a period of time I instructed my people to pay the invoices. I could not hold up the payment any longer without some justification of wrongdoing. The only way I could conclude a problem and withhold payment would be with Doctor Mathenson's help and that was not going to happen."

Dobbs thought for a moment, "If you were to call the number on the canister, who would answer?"

"Doctor Mathenson. If she is in her office she will answer."

Dobbs picked up the inventory document that had been entered into evidence and looked it over. He then requested Judge Thompson to allow the witness to review the invoice. Thompson agreed.

"Ms. Hinson, could you please review the invoice I just handed you, and from the information contained on the document can you tell me the contents of the canister in question?"

Hinson adjusted her glasses as she looked at the invoice. She then said, "As I testified to earlier, I have no official understanding of the canister's contents, from either the invoice or from Doctor Mathenson."

Dobbs then asked, "Let me ask you this, do you have any idea what may have been inside the canister?"

Downing quickly objected to the question, "Your Honor, the witness has in her own words stated she is not a scientist. Any answer she may give would be clearly only her opinion. Perhaps even rumors or watercooler hearsay that she may have come in contact within her own workplace. Office gossip. Surely this is not admissible evidence."

Thompson thought for a moment, then he said, "Sustain. Mr. Dobbs, your question asks the witness to answer a question she has already stated by her own admissions she was unaware of the canister's contents."

Dobbs breathed deeply, "Ms. Hinson, can you say by your experience, is the canister in question the type used to ship biological samples in a frozen state?"

Hinson responded, "Yes, I do know that these type of canisters are used to ship biological materials in and out of our labs."

Dobbs looked at the jury and paused, giving them time to absorb what Hinson had testified to. He wanted to make sure they fully absorbed her testimony. The jurors stared for a moment at Hinson and then three turned their attention toward Elizabeth.

After Dobbs had given the jury time to think about the question he asked, "I have one more question for you Ms. Hinson, what kind of person do you believe Doctor Mathenson to be?"

Downing quickly stood up and objected, "Your honor, once again the prosecution is asking for an opinion of the witness. As I previously stated, the witness's testimony could be tainted by many factors, not relevant to this case. Jurors have no way of knowing the basis for the witness's answer. I would ask you to prevent her from answering a purely opinionated question."

Thompson paused for a minute, "This case is about a lot of different elements. I think it is important that the jury hear from people who were dealing with Doctor Mathenson daily. They have an understanding of her relations with those she works with. The question of properness when it comes to the finances of this case are clearly tied to what is considered proper in medical research ethics. I am going to let Ms. Hinson give her opinion. With the caution, her answer should be confined to her professional working relationship with Doctor Mathenson."

He looked clearly at Ms. Hinson, "Is that understood?"

Hinson replied, "Yes your honor."

Dobbs paused before asking the question once again. "Ms. Hinson, do you have an opinion on Doctor Mathenson, relevant to your association in the course of your work?"

Hinson was for the first time clearly off her game. She was prepared to give clear testimony concerning the question of grant money. She was not prepared to give an opinion of her personal assessment of Mathenson in open court. Especially with her sitting twenty feet away looking at her.

"Doctor Mathenson is one of the most dedicated people I am associated with. She is smart, driven and totally dedicated to her work...which she believes is the only mission the university has. She fails on every account to understand that there is much more work and research being done at the university by others.

She treats me and those not in her inner circle with total disregard for our job, responsibility or our feelings. She fails to understand that her job as head of the genetics' department is to not only direct research, but to be the bridge between the projects funding and my department. If there is any question of proper expenditures, it has to be justified by my people in their accounting process. Each week the department accountants submit to me a list of problem expenditures for all grant projects that need to be justified. That weekly report contains more discrepancies from her department than any other. It gives me the impression that she could care less about the accountability of funds. When I raised the issue with my superiors, they tell me.to "fix it." Yet, having said all that, I believe she has no alternative motives, other than to find the answers to the scientific questions her lab is seeking. If she was to trade jobs with me for one month, she would have a better understanding of the complexities of keeping her department funded properly, especially with regards to federal grant monies."

Dobbs glanced over his notes. After a few moments he looked at the judge and said, "At this time, the prosecution has no further questions."

As Dobbs spoke to the judge, Charlie slipped Downing a note instructing him not to press her much further, except to ask the questions he listed on the paper.

As Dobbs returned to the prosecution's table, Downing hesitated for a moment before standing. The courtroom was noticeably quiet. After a moment, Judge Thompson asked, "Mr. Downing, do you need a recess before continuing with the questioning of this witness?"

Downing stood and said, "No, your honor, I am ready to proceed. Ms. Hinson, you have been the controller for Doctor Mathenson for twenty years. Is that correct?"

"Yes, that's about right. I also oversee other project assets, but primarily Doctor Mathenson's project. As it has grown in size and financial complexities in the last ten years, it has become my primary responsibility."

Downing rubbed his face as he laid the small piece of paper back onto the defense table. 'Ok, you say twenty years. Before I get back to that question, let me ask you another. I see you are wearing a wedding ring. How long have you been married?"

Dobbs stood and objected, "Your honor, in what way can how long Ms. Hinson has been married have to do with her testimony here today?"

Thompson gave Downing a questioning look, "Well…is this one of your theatrics that I warned you about this morning? If it is, I warn you Mr. Downing, I will not be happy with your question."

Downing stepped toward the bench, "Your honor, Ms. Hinson has stated in her testimony she prides herself in her professional dealings with the doctor's research and grant monies. She has also testified that for twenty years she has overseen that relationship. I am merely trying to establish that she exercises the same strict guidelines in her personal life as she professes in her professional life. If you would give me a bit of latitude I can move my question along pretty quickly and make my point."

Thompson stared at Downing with a stern look. Finally he said, "Wherever you are going, get there. Is that understood?" He looked at Dobbs, "Over ruled…for now."

Downing again looked at Hinson, "How long have you been married?"

Hinson with a bit of a disgusted tone to her voice, answered, "Twenty-three years."

Downing again walked to his table and looked at the paper on the desk before turning to Hinson, "Does your husband write checks?"

Dobbs stood and objected to Downing's question. Thompson looked at Dobbs, "Sit down I want to see where this is going. And you, Mr. Downing, get to the point."

Downing asked again the question, "Does your husband write checks?"

Hinson sat composed once again as she answered, "Yes… we both write checks. But, I can't see where this…"

Downing stopped her as he said, "Do you or your husband balance the checkbook?"

Thompson was clearly intrigued by the question as he leaned toward the witness.

Hinson responded, "I balance the checkbook. After all, I am the accountant in the family. It comes naturally to me."

Her answer drew a few chuckles in the courtroom.

Downing then asked, "Does your husband ever forget to write down in the checkbook what he wrote a check for?"

Hinson wrinkled her face with confusion at Downing's question as she answered. "Can't see where that has anything to do with this case. I'll not answer that question."

Downing looked at the judge, "Sir, could you please direct the witness to answer and remind her she is under oath."

Judge Thompson leaned forward as he said, "Ms. Hinson you will answer the question put to you by the defense counsel… is that clear?"

Hinson's movements in the witness chair clearly showed she was out of her element with the question. "My husband…my husband sometimes does forget to complete the information. But, he never writes a check for funds not in our account! I balance the checking account and everything is fine after I justify the funds. We have never had an overdraft in our checking account."

Downing shook his head as he said, "I would suspect many a good wife keeps the checkbook balanced and up to date. So given that even in your own life the money and documentation is not always the way you want, you manage to keep the books in balance. Wouldn't you agree?"

"Yes, I would agree."

Downing then looked at the jury to make sure they were paying attention as he said. "So I would be correct in saying, that whether it is the funding of a research project or in your personal finances, you know that people do not always complete their documentation. And that lack of documentation does not mean there is intent to mislead or misrepresent the funds, it just means that people who are not diligent in finances tend to pay little attention until the books need balancing or there is a problem. Is that correct?"

Hinson paused and raised her head in a proud stance, "Yes, I guess I would say that would be correct in my experience."

Downing quickly responded. "Then it is possible that Doctor Mathenson's actions were not an attempt to misdirect the project money, as it was her poor understanding of the importance to your department in keeping the money straight. A job that you pride yourself on doing correctly, just like in your own life?"

Hinson hesitated for a moment, "I guess that could be correct in this case."

Downing turned to her, "Ms. Hinson, is that a yes or no so the jury clearly understands your answer in this matter?"

Hinson, "Yes!"

Downing shook his head in agreement with her direct answer. "When the prosecution was questioning you, if you would recall." Downing paused and picked up the piece of paper that

Charlie had handed him. "Yes here it is, you said, you were instructed to "fix it" by your superiors, is that correct?"

Hinson thought about her earlier answer, "Yes that is what I said. But that meant… They…"

Downing stopped her and looked coldly into her eyes, "Then it is correct in saying her lack of actions with the money was well known in the department? And you, Ms. Hinson, made the final call to give her a pass in her poor bookkeeping skills?"

Hinson hesitated as she said, "Yes, when you put it that way, but I made professional judgment calls to keep the program going."

Downing quickly asked as he looked toward the jury. "Why would the very people who are in charge of her funding, permit you to allow her such irresponsible behavior with project funding? Are they also the same people who dismissed her and illegally withheld her tenure from her job? The same people who I assume turned her actions over to the Justice Department for investigation. The same people who sent you into this courtroom to testify about her poor accounting skills, and by your own testimony was allowed by you when directed to, "fix It". Isn't that correct?"

Hinson leaned forward eager to answer his question, "Yes and no. Neither my superiors nor I contacted the Justice Department. My supervisors were not unhappy with her research work in genetic medicine. As to the money questions, they expected me to fix any problems with the money, as long as it was not an illegal problem. Is that uncommon? Yes. Is it unheard of for the controller to take whatever actions that are necessary to correct problems? No! Doctor Mathenson's program brings in hundreds of applications for postdoctoral positions every semester. Some of the very best young minds enter the world of research and want to be part of this program.

Each month companies donate scientific equipment and materials to the programs. Doctor Mathenson's research is, or was, considered one of the top in this country or even in the world. It is a tremendous benefit to the university to have her on their staff from many different standpoints. That was up to a point."

Downing looked at the jury and raised his hands indicating he didn't understand Hinson's answer. Then he said to her. "The university has no problem with Doctor Mathenson's work or her use of funding. Even with all her shortcomings in reporting protocols for money and her research? The same Doctor Mathenson they dismissed from her life's work that is famous in the world scientific community?"

Hinson gave Downing a wondering look. She was not sure if he was making a statement or wanted an answer from his rambling statement.

Downing stood looking at the jury not saying anything for what seemed like a long time. Finally he stopped his empty stare and looked at Ms. Hinson, "Let me ask you a question that requires your opinion. Why is Doctor Mathenson on trial?"

Dobbs stood and objected loudly, "Your honor this witness is not here to help the defense understand why Doctor Mathenson is on trial. Any personal opinion would be irrelevant. She is here to give expert testimony about the grant program's finances in regards to Doctor Mathenson's funding."

Thompson looked down at Dobbs as he said, "Counselor, I am going to allow the witness to answer. I want the jury to know what she thinks. In your earlier questions to her, you established she is allowed to have thoughts about this case, supported by only her opinion. Your objection is overruled, continue Mr. Downing."

Downing did not hesitate as he asked once again, "Why is Doctor Mathenson on trial here today?"

Hinson sat back in her chair and thought for a moment before answering. "In my opinion, I believe she is not on trial for her poor business skills or even her ground breaking genetic research. I believe this is about who controls the research into new biological areas. I think she has taken her own personal research goals to places some believe she should not have gone. I don't know what she has discovered, but I do hear the other side of the conversation, a conversation whispered off the records. That whisper says she has broken the Laws of God."

Dobbs again stood and protested her comments. "Your honor, the government has charged Doctor Mathenson of embezzlement, under section 18 of the federal law. Not breaking the Laws of God. I have read the legal text and statutes, it says nothing about, Law of God in the embezzlement section of the statute. Ms. Hinson is the prosecution's witness. The defense has maneuvered her into introducing another element into this case. An element of faith, not legal statutes."

Before the judge could rule, Downing stepped back towards the defense table as he said, "Your honor, we have no further questions for this witness. But, I would like to have her remain available." He glanced out of the corner of his eye toward Dobbs. "In the case that her testimony on the Law of God may be needed at a later time."

Thompson looked at Dobbs, "Does the government have anything further for this witness at this time?"

Dobbs was in a momentary conference with Ms. Williams. After a moment he stood and said, "No, your honor, at this time we haven't anything further. But, I would ask any testimony referring to the Laws of God be stricken from the record and the jury advised to disregard that term."

Thompson looked at the clock on the back wall, it showed quarter till six. "I will give the question consideration. But due to the lateness of the hour, I recess the court until tomorrow morning at ten thirty. Does either council have anything further before I adjourn today's proceedings?"

Dobbs and Downing stood and collectively said, "No your honor."

With that answer, Judge Thompson instructed the bailiff to remove the jury from the room. After the last member had exited, he struck his gavel and the bailiff announced, "All rise, court is in recess." The courtroom began to empty as the judge exited the room.

Just as the door closed to the judge's chambers a voice Charlie knew called out. "Elizabeth Mathenson, you have heard the word, you have broken the Law of God. Just as in the time of Jesus a Babylon Whore will damn the earth. You have risen once again to try and damn mankind. But I will not allow that to happen."

Before he could say anything else the U.S. Marshals pulled him from the courtroom. Charlie could hear his voice fade as they removed him from the building. Elizabeth sat down and held her head in her hands. "Why is that man tormenting me? I have done nothing to him. I just don't understand why he is here."

Downing sat down beside her, 'Elizabeth, don't let him bother you with his words of damnation. He is only looking for the publicity for his Church in the Sky. Those that know him realize he is just a loud mouth man who wants to believe he speaks for God. He only speaks for himself."

Charlie looked at the stunned group. "We can't let him distract us. We are nearing our turn in this case. Cain can say what he wants, but he doesn't sit on the jury. We have to concentrate on tomorrow. So far we have countered their witnesses' testimony by making each juror think about both sides of this case through their own eyes. You have made them think about keeping their own money straight in their checkbooks. Will they equate that personal experience to the controller's ledgers in Elizabeth's case? We can only hope each of them have missed recording a check, or tried to balance the checking account when someone else did not record their checks. It may give them an understanding that they had no illegal intent in their actions. Is Elizabeth's lack of concern or diligence in keeping her paperwork up to date any different than their own mistake? So far they have not shown any fraudulent use of the money. They have shown open questions concerning why and who approved money spent in the Korean research project associated with the woolly mammoth project. They have also raised questions about other large expenditures that she made in Israel and England without proper documentation. Dobbs has done a good job showing she is a poor bookkeeper and has a lack of empathy for those trying to keep the money straight for her projects. You noticed members of the jury kept looking at Elizabeth and then towards Hinson during her testimony. Each time, one of the jurors is wanting Elizabeth to answer the prosecution's questions they have raised. So far the jury seems to have more unanswered questions than your cross examination has answered. We still have a lot of work to do convincing the jury that her intent was not criminal in nature. And as to the question of her personal attitude toward others, we can only hope when she takes the stand, they will form an opinion during her testimony." He looked at Elizabeth. "A positive opinion. Do you understand how important it is to have the jury think of you in a good light?"

She answered, "Yes."

He then laid his hand on her shoulder, "and you need to forget about Jedidiah Cain and his Old Testament ravings."

Charlie smiled at Elizabeth as she wiped away a tear. He opened his briefcase and slid his papers inside as he said to

Elizabeth, "Time has come to have a real conversation about your work. That last witness both helped and hurt you. But, her answer as to why you are charged now has to be dealt with in open court. Every juror is asking themselves, 'what does it mean, she has broken the Laws of God?' If they heard outside this courtroom about you creating life, they may now believe even people who work with you believe you have done this thing. You'll have to answer that question. If you don't, they may hear the words of Cain in his ravings. We can defend you on the legal questions, but Hinson's testimony went somewhere I had not anticipated from one of the prosecution's witnesses. And Cain is fanning the false idea of hatred of you. I knew sooner or later the ethical questions would come into play, but I had hoped we could introduce those questions with our witness. We do know by Hinson's testimony that the university knows more than they are saying about the case."

Charlie then looked at Downing, "I am sorry, I should have anticipated Hinson would raise the question of ethics and religion. That damned hilltop preacher and his crowd have been pushing his religious agenda for a while now. He got his information from someone that told him your work somehow is anti-God. Someone is feeding him bits of information, just enough to keep fanning the false impression of her into wild controversy. I broke my own first rule, never ask a question you're not prepared to have answered in front of the jury. We now must answer that question. Clearly and so the jury understands why she said that."

He looked at Elizabeth, "Before we can convince the jury about your case concerning the money and now the God question, it is time you answer it for your own defense team. You have been holding back tying all the points together. No more half answers."

For the moment, as the courtroom cleared, no one said a word about what Charlie had just said to Elizabeth in a not so subtle tone of voice. Then Tucker, who had been listening, pointed toward the doors of the courtroom. "There, he can help answer your questions. Maybe better than anyone else. Angus Whitechurch is one of the greatest scholars of theology I have known and he is also a man of science. A combination that has caused him great personal turmoil as he searches for answers in both worlds."

Tucker and Elizabeth quickly made their way to where Angus was standing. As Tucker approached he said. "I thought your plane was not going to arrive until later this evening?"

Angus, who appeared tired, sat down in the back row of the courtroom. "I called on an old friend at the airlines to see if he could get me on the early flight. Jet lag is not good for an old man. Thank goodness he managed to get me a boarding pass and it was an upgrade to first class. Nice flying not all crowded together. I managed to get some sleep on the way."

About then Elizabeth put her arms around him and began crying, "Oh Angus, I am glad you are here. I am in so much trouble and all of our work is going to be lost in the legal system I'm afraid."

"Now Elizabeth, let's not get all emotional. Let me get up to date on where we are and Charlie can decide if I can help your case, or hurt it."

Charlie stepped close and offered his hand to Angus, "Good to see you again. I am somewhat surprised you agreed to be part of this. Your government and Stonemore may not like you getting involved in an American problem, especially one that has become much more political and now poses some theological questions."

Angus shook his hand, "Well Charlie, you see that is the problem. Elizabeth and I have spent many hours pondering… whose problem is this? We believe it is the world's problem and time has come to see if we can help fashion an answer that people on both sides can live with. But enough of this for now. I need some supper and a drink."

Tucker gave Angus a quick look as he made the comment about a drink.

Angus patted him on the shoulder, "Don't worry my old friend. I meant a cup of tea. I need to keep my wits about me in this foreign country. After all, your President may see me as an undesirable and deport me."

Charlie was still absorbing Angus's answer that Elizabeth and he had knowledge of the problem. Then he asked, "You make it sound like you two knew this was coming. I don't understand."

Tucker stepped up before Angus could answer Charlie's question and said, "Welcome my old friend. Charlie was just telling Elizabeth it is time she explains what is going on. And I said you were the best person in the world to help everyone understand what's happening with her case."

Charlie turned and looked at all three of them with a strange look on this face as if he were the only one of them who did not know the full truth. He then said to them, "We have the small conference room allotted to us back at the hotel this evening. We'll get some dinner and then sit down and you three can explain what is going on. I am tired of feeling like the only one in the room who doesn't know. The prosecution will put the head of building security on the stand first thing in the morning. He'll introduce the tunnel video evidence. He has not added any more witnesses onto his original list. Dobbs has dropped the second controller's testifying. Hinson covered everything he wanted into evidence. Her testimony gave the prosecution a few problems, but nothing Dobbs won't give his spin to when he gives his closing argument to the jury. He'll not take a chance to have conflicting answers from two controllers from the same office. If he needs to, he will recall Hinson to clarify any missed answers, but I think he'll clean up her testimony in closing. He'll close the prosecution's case with Clay Gordon's testimony. Not real sure what Clay is going to testify to at this point. Whether he is acting on the part of the university or is he going to reveal his government's interest in this case. If he does, he knows we will reveal on cross examination that Clay is passing information to his government. That's a big risk for both Clay and Dobb's case. Whatever he does on the stand, it will be our turn after Clay testifies. We can at that point decide if we want to confuse the jury with a foreign spy in Elizabeth's case. Sometime in the afternoon we may be calling our first witness if the hour is not too late. From this point on, I will be handling the case. Downing's done a good job so far on cross with Dobb's witnesses. So far I don't think he has shown anything damaging to the jury other than her poor bookkeeping skills. But this question of God's Law has put a whole new meaning to this case."

He looked at Angus, "My friend…I have a lot of questions for you and as to how we move forward and not let the science overwhelm the legal charges against her."

Elizabeth then asked, "Who is going to be our first witness?"

Charlie said, "We'll discuss that this evening, after you and Angus explain to all of us what is going on. No more half answers. You want to stay out of jail and get a shot at going back to your research, you have to explain what you have done. By morning all of us should know as much as you and Angus know about the science and the money. Otherwise if you continue on this course of actions, you'll be taking your chances with the jury. At this point I don't think you want to do that. Now, let's get some dinner and then meet at the conference room at seven thirty." With that final word the group exited the courtroom.

Chapter 25
The Knowing

Jake was finishing adjusting his tie as the elevator doors opened. He looked at the clock on the wall behind the check in counter. It showed five-thirty. Earlier he saw no activity in the lobby as he made his way to the dining room hoping to find coffee waiting. He knew it was going to be a long day in the courtroom, especially after spending four hours hearing what Elizabeth and Angus had to tell them about so many things. After the late evening conference he had spent a few more hours researching some of the things they had talked about. He knew Charlie would want the background information before he began questioning the defense witnesses.

As he stepped into the dining room he saw sitting in a booth toward the back, Charlie. He recognized his snow white hair and suspenders from behind. He thought he was going to slip in and get coffee and not see anyone at this early hour. As he stepped close to his old friend, Charlie turned his face towards Jake. Even in the dim light Jake could tell his old friend had tears on his cheeks.

"Charlie, is there something wrong? Can I be of any help?"

Charlie, who had been surprised by Jake's sudden appearance, realized his appearance as he wiped his face with a napkin. He then said, "No Jake…nothing's wrong that either of us can fix. Sit down and give me someone to talk with."

Jake sat down slowly realizing something was wrong. He had known Charlie for years, but he had never seen him cry, even when his family died."

Charlie sat for a moment and held the coffee cup in his hands as if he were warming them. Finally he said, "After last night I can't stop thinking about Rose and Timmy and how much I miss them. I never thought of myself as a religious man or even a scholar of science. I built my life on the law, words that I could see and read. Put my fingers on the pages and feel the ink. I trusted what the law said. I knew it was not always perfect, but I could depend on the law being there. Just right and wrong. No question of is he guilty because of the words in the Bible or science book. Justice by the written word of man. But now I don't know. If I had to choose science, I think of my family. If I think of the Bible I have to have faith, something I've never had. I wish that I had never let myself get involved in this case. Why couldn't they have just left me in the mountains fishing?"

Neither man said anything for a few minutes. Finally after a time Jake spoke as he stared at his coffee cup. "When they first told us that Ella had cancer, the air went out of my world. I can't imagine what it must have felt like for her. It was just a small pain that would not go away. She figured it must have been a pulled muscle from tending her garden. At first she never said anything to me. But after a couple of weeks she went to our doctor. They ran some tests, told us it was more of a precaution than anything else. Probably nothing to worry about. Two days later they called and told her the doctor wanted to talk with us. She knew at that moment. I figured they could fix anything. We had good insurance and there were some of the very best hospitals in the country just a few miles down the road in Baltimore. I could tell from the look on the doctor's face, it was not good. He told us he had gotten her into a drug study with a promising new drug therapy. When we went down for the first treatment I felt like she was a rat in a lab. They were nice enough and they explained the many side effects. But, they also said her type of cancer responded well to the therapy. We would have done anything to fight the cancer. But each day I watched as she died a little more. Finally she asked me to let her go. She had lost so much weight it was hard to recognize her. The day she asked me to carry her down to the end of the dock and put a fishing line into the water. She wanted to catch one more crappie. She closed her eyes waiting for a fish to bite. I sat there holding her until her line bounced and I pulled in a small fish. I whispered to her the crappie had returned. I told myself she heard me. It took a month for her to die. I am still dying inside. But, what I heard in the room last night makes me angry. Why couldn't science have found at least some of these answers sooner? And if they had, would the drug companies and big businesses allow it to be given to those who need it. And what about the government? They take years to approve new therapies. How many more are suffering because of the lack of answers? Then you pull me into a case of embezzlement and low and behold I find out what Elizabeth's research has discovered. Could it have made a difference, I don't know? Could what you are wondering come to pass with Rose and Timmy? I don't know. But this is not going to end when this trial is over. Everything will change."

Charlie stared at Jake before he said, "I am sorry I got us into this. But, I have a client that needs defending. What she has done has nothing to do with the rest of this. But her guilt or innocence will depend on how the jury understands what she says in her explanation of what others have testified to. That explanation is very crucial as to how they judge her. Just look at us and how we responded to what we know. I am going to lay it out for the jury the best I can."

Jake smiled at Charlie, "For two old men who were once the pride of the Justice Department, we're sure a sad case for life after seeing the inside of the legal system."

Charlie chuckled at Jake's comment as he laid a ten dollar bill on the table. Then as Jake started to stand, he laid his hand on his sleeve. "Jake…do you believe in… God?"

Jake slowly sat down and shook his head a bit as he chuckled softly. "Do I believe in God?" He looked up at Charlie and smiled a bit. "I don't believe in the God Jedidiah Cain holds up like a carnival side show. But, when Ella was sick I prayed every day to the God she believed in. I asked him for her sake to make her better. I can't say I was a religious person for most of my life, but Ella truly was. She told me a long time ago when we talked about religion that she expected me to attend church with her on special occasions. You know, Easter and holidays. I always did, if we were not tied up in Washington prosecuting someone. In truth I went to satisfy her, not because I really believed. When she became sick, I realized all those years she tried to get me to see there was something bigger than ourselves, some of what she believed must have rubbed off. At first when she got sick I prayed she would respond to the experimental treatment. But near the end I found myself praying for her to find peace. She suffered so from the cancer. I believe she suffered even more from the treatments. I remember the last day she spoke to me. She was laying sideways in the bed and I was sitting next to her. She reached out and touched my face. She was so weak her voice was little more than a whisper as she said, "I think it is time to go my love, I can see the light." I held her hand and squeezed it as tears ran down my face. We held hands for nearly an hour. Each minute I felt her hold weaken. Finally I felt life pass from her hand's embrace. I never realized I could feel so much pain. In my grief I…" Jake stopped for a moment and reached into his suit pocket and brought out a handkerchief and wiped away a tear. He then looked up at Charlie. "You know what I did then? I thanked God. I thanked God. Not because she had died, but because he stopped her terrible pain. What kind of man does that make me? Thanking God for taking the only thing in my life I truly loved. Over the next couple of months I even thought about joining her. There was no joy or purpose left in my life. Then one warm spring morning, I walked down to the dock, a place where she loved to sit and read. In a bit of dirt between two boards, I saw a small yellow flower growing. Then I remembered Ella would tell me each spring of the flower that grew between the boards on her dock. Somehow Ella's flower was telling me life goes on. I can't say things changed greatly that day as to how I felt, but I do know I started to look beyond her suffering. I started to remember the things that made her so wonderful."

Jake hesitated for a moment before answering Charlie's question. "Do I believe in God? I want to. Ella believed and it gave her peace. And in a small yellow flower I found some of the peace for myself. But after listening to the conversation last night…I don't know."

Charlie finished the last drink of his coffee and stood. "We better see if the others are ready. I want to get to the courthouse early this morning."

Near eight o'clock Charlie and the others came up the street that ran in front of the courthouse. They could see the street was full of satellite television trucks. Dozens of people were moving around the plaza carrying signs. Most were anti-science. A few signs called Elizabeth the Whore of Babylon. Charlie was thankful they were not inside the courthouse. But, he also realized the jurors would see them when they were escorted into the courthouse.

Charlie pointed at the gathering just before they turned into the basement entrance. "You see…signs protesting Elizabeth's science and the label Jedidiah Cain gave her. Not one sign protesting that she is an embezzler. She is on trial for one thing and the world is judging her for another." Nothing more was said as they left the daylight of the street and entered the dark basement access to the courthouse.

Two hours passed as they prepared for the day's testimony. Last night they decided to change their strategy as to how they were going to handle the security video Dobbs was likely to enter this morning. The head of security was unlikely to add any damning information to the case. They also decided to wait and see as to how to handle Clay Gordon's testimony. If he revealed his link with the Israelis, they would have to go after him and challenge his testimony. If he stayed with his role as the university's legal representative, they would only address elements of his testimony against Elizabeth.

They also knew that her guilt or innocence would be decided by her testimony about the grant funding issues in her program. The jurors want to hear why she used the funds as she did. Did she realize there were questions on the direction her research was taking? Did she just ignore those that were telling her she was off course in her work? It would also have to be decided on the unsaid allegations against her research. Charlie knew the jury's curiosity had to be answered. The jurors were only human. He knew he could handle the issues dealing with the money. He was confident he could show the jurors that the charges against her were very thin and lacked solid evidence of intentional wrong doing. And he would try and make them understand, although her actions were negligent and poor in judgement, which is not necessarily a crime she was charged with. When it came to the question of her research and where it had taken her, Charlie needed Angus Whitechurch to lay it out in terms the jury could understand. He also knew Dobbs was going to go off the deep end at Angus's testimony. Charlie was relying on Judge Thompson allowing his testimony into evidence.

One big question would also have to be answered, what was in the cryo-chamber and where is it now? Especially after Dobbs enters the building's security footage into evidence. It is the one piece of solid evidence that is damming to Elizabeth. Dobbs will try to convince the jury it was removed by Elizabeth. He would tell them she was the only person who had access inside the lab to the container. In last night's conversations with

Elizabeth, Charlie learned the truth of what was in the canister. Now comes the tricky part, revealing it to the jury, so they understand why she removed it. He hoped the knowing of all the facts would give the jurors reason to try to understand her intentions.

Time passed quickly and before Charlie realized, it was nearly ten-thirty. The court bailiff told both counsels it was time for the judge to return to the courtroom.

At ten-thirty, by the courtroom clock, Judge Thompson opened the door and entered the room. He adjusted the papers on his desk and made a few notes. Then he looked at both counsels. "Gentlemen before we proceed with today's agenda, there was an objection raised by Mr. Dobbs yesterday concerning the term, Laws of God. After review of the testimony and the context of the day's events, I have decided to allow Mr. Hinson's statement to stand. Now, are we ready to proceed?"

Both counsels stood and said at almost the same time, "Yes your honor."

Thompson looked at Dobbs, "Good, call your next witness."

Dobbs replied, "The prosecution calls Patrick Mellot to the stand."

A thin man in a dark ill-fitting suit made his way to the witness stand. He looked as though the suit was made for a man one full size larger than he was.

Dobbs approached after he took the oath. "Can you please state for the recorder your full name?"

"Yes, my name is Patrick Mellot."

"Mr. Mellot, can you tell us what your job is?"

The man adjusted the way he was seated as he said, "I am head of security for the University Genetics Lab building."

"And how long have you held this position?"

"I have been with the university security division since 1995. For the last ten years I have been head of security operations inside the building."

"Do you know Doctor Mathenson?"

"Yes, I have seen her on many occasions. I believe she has been working there even longer than I have."

"In the course of your job have you ever had a conversation with her?"

"No…I don't think so. I see her when she enters the building and we may exchange good mornings. But most of the time her mind is elsewhere and she takes no notice of anyone around her. I think she is a workaholic. As far as I know, she hasn't intermingled with any of the building's staff for as long as I have known her."

"So would you say she is unaware of the security protocols in the building?"

Charlie stood, I object your honor. Mr. Dobbs is placing descriptive words into the witness's mouth."

Thompson looked at Dobbs, "Please restate your question."

Thompson thought for a moment, "How would you describe, in your professional knowledge, Doctor Mathenson's understanding of the building security?"

Mellot was still thinking about why he could not answer the first question when it struck him the court was waiting for his response. "Uh… well I guess I would say she must know it's there because she shows her ID card to the guard at the entrance when she comes in. But, she may do it while reading papers in her hand. She is like a lot of others, they just take us for granted and pay no attention to us. We're like the furniture in the lobby."

"Ok, let's talk about the security video system in regards to the access tunnel under the building. I am going to show you a video that your department downloaded for me in response to a warrant for evidence in this case. Please explain to the jury what they are seeing. The piece of video is from June 9 of last year."

The court bailiff pulled down a screen and in a moment an image appeared. It showed a high view of an empty hallway that ended with two metal doors. Mr. Mellot began, "Well, this is the lower access hallway. By the time code in the corner it looks to be twenty three hundred hours and ten minutes. That's ten minutes after eleven at night. The doors you are looking at open into the boiler room that leads to the old hospital basement. Now days it is only used for moving large pieces of equipment in and out of the building." He looked back at Dobbs with a look of asking what else.

Dobbs walked closer to the screen and said, "Wait… in just a moment two people are going to enter the screen. What can you tell me about them?"

One minute passed and two people in black hoodies enter the video image. The larger of the two was pushing a two-wheeled cart. "Mr. Mellot, what do you see here in the image?"

"Well, those two people are removing what appears to be a cryogenic canister from the lab building."

He pointed toward the screen and says, "You see that placard on the side? That means it has biological material inside and it poses a danger of extreme cold if it comes in contact with skin. They have lots of them in the storage room in the basement of the main lab."

Dobbs stopped the video so he could point to what Mellot was talking about. He then turned to Mellot. "Can you tell me what this canister has in it and who it belongs to?"

Mellot squinted his eyes as he looked closer. "Not real sure, it's kind of blurry, but I think after talking with your investigators, it must be the canister you all are looking for."

Dobbs then moved the video ahead frame by frame until he pointed toward the glass in the door. "There, in the glass. The reflection of the smaller person. Can you identify the person? Is it a him or her?"

"Well… not sure, but based on what your people said. I am thinking maybe its Doctor Mathenson. Not real sure. The dark rimmed glasses kinda' looks like hers."

Dobbs stopped the video and handed Mellot a couple of papers. "Can you tell me if Doctor Mathenson was logged into the building's security system that evening?"

Mellot took the paper and said, "Looking at the print out of those who accessed the building that evening, it looks like she was in the building."

Dobbs took the paper back from Mellot and laid it on the evidence table. "So to be sure the jury understands, Doctor Mathenson was in the building. And by your testimony, she was seen leaving the building with another person at ten minutes after eleven removing a cryogenic canister. Is that your testimony?"

Mellot paused for a moment and then said, "Yea, I reckon that's about right. I wasn't on duty, but from the video and the records you showed me, I would say that is about right."

Dobbs abruptly said, "The prosecutions has no further questions for this witness."

Thompson looked over at Downing, "Your honor, my co-counsel Mr. Morgan will be handling the case from this point forward."

Thompson looked at Charlie, "Well, let's get about your cross Mr. Morgan."

Charlie walked to the middle of the room and looked at the jury. "You say you have worked in the genetic building since 1995, is that correct?"

"Yes, sir, that's right"

"In all those years how many times has someone used the access tunnel for things not logged into your security records?"

Mellot chuckled, "More times than I can count. You see, those tunnels are the kind of a place where college boys can have late night meetings with girls. There are some that try and gain access to the old hospital, thinking there may still be some drugs inside. Ain't been any there since, oh, about 2000."

"Isn't it correct that in other security videos that your department gave to the defense as evidence, it shows other illegal activities, such as theft of large quantities of dry ice and toilet paper?"

Again Mellot chuckled, "Yea, them kids can figure some ways to get into the tunnel and access the dry ice storage room and the janitor supply room. They take the dry ice for Halloween parties or just keepin' beer cold for a frat party. Three cases of toilet paper were stolen and ended up in the trees in front of the administration building. Students were protesting a new drinking prohibition on campus. It didn't work."

"So, it would be fair to say these are really not secure tunnels at all?"

"Yea, I guess you could say that, but we try and keep them darn kids out of the lower levels. With just two of us on duty at night it is near impossible to be everywhere."

"Okay, Mr. Mellot. Did you recognize Doctor Mathenson's face in the reflection before one of the prosecutions investigators asked you if it could be her by name?"

Mellot thought for a moment, "Truth is, until he played it back, frame by frame, I never even saw a face. Then he said, 'Look real close at the image in the reflection and didn't it look like Doctor Mathenson?' I thought, well maybe it was her. It did kinda' look like her dark rimmed glasses. I couldn't really say for sure it was her face, but maybe the same type of glasses as she wore."

As he spoke about her glasses the jury looked toward Elizebeth to see for themselves. She was wearing delicate glasses with no visible frames in court.

Charlie stepped toward the witness and said, "One last question, did anyone report a stolen cryogenic chamber to your security people."

"No, not really. Those kids have been known to take stainless steel containers and cut them in half. They make beer coolers out of them. I'm guessing, if we got a report, we just figured this was another canister on its way to becoming a frat house beer cooler."

Charlie looked at Thompson, "We have no further questions for this witness."

Thompson dismissed the witness and then said to Dobbs, "Are you ready to call your next witness?"

"Yes your honor, we would like to call to the stand, Mr. Clay Gordon."

Clay made his way to the witness stand and took the oath. Dobbs approached him, Mr. Gordon, can you tell us your full name.

"Clay Doran Gordon"

"Mr. Gordon, what is your profession?"

"I am the senior professor of law at the university. I have been on the faculty since 1981."

"Do you know Doctor Mathenson?"

"Yes, I have known the doctor for as long as I can remember. I am not real sure when we became acquainted, but it has been many years."

"Would you call her a friend?"

"Yes."

Dobbs paused before asking the next question. "Has the doctor ever asked any questions that were of a legal nature concerning her work?"

Gordon looked over at Elizabeth, "Yes, last summer at a faculty dinner she told me early in the evening she needed my advice on a hypothetical legal question."

Dobbs waited for him to go further with his answer. Gordon again looked at Elizabeth, "She wanted to know if there was any way around the university not taking possession of a discovery not found in the direct line of her work."

Dobbs looked at the jury with an inquisitive look on his face. He them asked, "Why would she want to know about something like that?"

"She knew I specialized in teaching patent law, especially international patent law. I teach basic law, but I have become known over the years for advising corporations and investors on how to navigate the complicated world of patents. Doctor Mathenson knows, like anyone who works for most companies or universities that if in the course of their work a discovery is made that can be patented, the ownership goes to the university or company for whom they are employed at the time. The person or group that makes the discovery will receive a percent of the possible income of the discovery. The percentages vary with different groups. To be honest, it was only a hypothetical question. I receive them more than you would think. People think they are not obligated to pay me if it is just a hypothetical question. It is not uncommon for a researcher to work for many years on finding ways to do things better. Their scope of attention has been totally on solving a problem and creating a new invention or medical cure. Then one day they find that answer. The next day, they realize who owns their discovery after years and years of work. They receive a small sum for their efforts. The university could in some cases receive millions. Especially when it comes to medicine."

Dobbs turned back to Gordon and asked, "In the case of Doctor Mathenson, what did you think when she asked you this hypothetical question?"

"I figured she was indirectly referring to her research and that she must have produced a discovery she wanted to hold on to. I figured it must be a major discovery to cause her to ask the question. I knew from the past that her lab had advanced the science of medicine several times with important discoveries. And each of those times she never questioned the discovery's ownership. But this time she did, at least hypothetically speaking"

Dobbs hesitated before asking his next question as he gave the jury time to absorb Gordon's answer. "Ok, now let me understand, Doctor Mathenson asked you about patent ownership, then later on she was terminated from her position and her lab closed for the time being. Did her conversation with you have any bearings on these events?"

"No, absolutely not. Whatever happened was not brought about by me." Gordon raised his hands as if in disbelief he was asked such a question.

"Then you're telling me that even after she revealed to you that she wanted to keep a discovery from the university, that you did not report her request?"

Dobbs made quotation marks in the air with his fingers as he said, "Even hypothetically speaking?"

Gordon paused and looked Dobbs in the eyes as he responded, "As an attorney, you and I both know when someone asks a question in a hypothetical manner, that is not an admission of anything. Before I could or would pursue Elizabeth for her question, I would have to have more information, much more information. I made a choice… I let her work continue without question, based on her passing question at an informal faculty meeting. If something changed in the future, I would have to reassess my earlier decision. I teach law. I make sure my students understand that when they work with the law, there may be times you may not like what rights the law gives to a client, and on the other hand, what rights it may take away. But using good judgement and researching the problem before you act will always keep you close to the truth. I chose to believe Elizabeth was staying within the written laws. Her past had given me no reason to think otherwise."

Dobbs asked questioningly, "Are there any other type of laws you have encountered in your many years of being part of the legal system?"

Gordon paused and looked at the jury, "Yes, I would assume you are referring to the laws of God. They are also written down for man to know and understand."

Dobbs looked at Charlie as he asked his next question, "You say the laws of God are written down. Are you speaking of the Bible?"

Gordon, again wanting to make a connection with the jury, looked toward the group as he answered. "They are written down in many places. If you are a Christian, they are in the Bible; a Muslim or follower of Islam, then you'll find the words in the Quran', and if you are Jewish, then the words are written in the Torah or Hebrew Bible. Each of the world's many religions has a set of written words that help them in their faith. A way to pass the laws from generation to generation. And when I say the law of God, I'm referring to any place the words of faith are written."

Dobbs walked to the prosecution's table and picked up two books. He then returned to where Gordon was sitting. He held up the first book. "Can you identify this book?"

Charlie stood and said, "Your honor, is the prosecution entering the Bible as evidence in this case? I would remind the court she is on trial for crimes involving funding of her projects."

Thompson looked at Dobbs. "I will have to agree with the defense, you have so far not laid the groundwork for this approach to this case."

Dobbs responded quickly, "Your honor, the term, laws of God has already come up in the trial and you allowed them to be entered into the record. I am merely asking the witness if he ever considers the words of religion when teaching his students the written legal laws. I believe this is an important part of the prosecution's case. This question is in anticipation of issues that will be raised by the defense team."

Thompson thought for a moment. Then he called Charlie and Dobbs close to his bench. He covered the microphone so the jury could not here his comments. In a whispered tone he said, "Charlie, I am going to allow this into testimony. That is unless you tell me you're not going to touch on this subject yourself with the defense witness."

Charlie looked at Dobbs and then back at the judge. "My client's testimony will be about the funding questions and the science of her work. I intend to try, try mind you, and keep the question of science from being turned into a religious question. But I know that may be difficult already. It was in my cross that the laws of God were first raised by one of the prosecutions witnesses. If Mr. Dobbs is intending to establish that legal laws must be separated from religious beliefs when establishing guilt, I have no problem."

The judge looked at Dobbs wanting an answer, "Well, is that your intent?"

"I have no intention of entering that place between the two. My only effort is to establish that both can sometimes be considered in life. But, when someone's guilt is proven by the legal system, it must prevail without influence from religion or any other beliefs. Hiding behind the Bible has no place in this courtroom by my hand." All agreed on his explanation.

Dobbs returned to Gordon and held the Bible up to him to identify, "What is this book?"

Gordon craned his neck a bit to clearly see the cover. "That appears to be a Christian Bible."

Then Dobbs held up a smaller book for Gordon to see, "And this book, can you identify it?"

Gordon sat forward once again to see the cover of the book before answering, 'Well… it appears to be a book about Charles Darwin and evolution."

Dobbs slowly walked back to his table and laid each book down separately, looking at the covers as he did. He then turned to Gordon. "Do you teach from either of those books in your classes?"

Gordon paused for a moment before saying, "I would have to say no, as a general rule, but the question has over the years sometimes come up in my law classes. I try to explain that the laws of man are written to be fair in the judgement of the person accused, assumes no guilt or innocence. I have also told my students those laws were written by those who often believed in a higher power of fairness and honesty. That higher power…I guess you could say it comes from a belief in some spiritual guiding force. Right and wrong. Some would say that means God. Others would say it simply refers to mankind striving to be civilized in its judgment of others. Without the written laws put down by man, we would be no more than the wild animals of the forest. But we have the ability to think and strive to govern our lives by just rules."

Dobbs seemed for a moment to stare past Gordon as he asked, "Which do you believe is correct?"

Gordon without hesitation said, "My faith is based in the belief that we found our way out of the desert by the commandment of God. Everything after that comes from those laws set forth by his commandments. As we have grown as a civilization, we have also had to grow in our laws. But the foundation for all law comes from the Commandments of God."

Dobbs turned and looked at the jury. "Do you believe Doctor Mathenson has broken the laws of your God or has she broken the laws created by man?"

Charlie stood and protested the question. "Your honor, we can line up a list of witnesses that would proclaiming my client is innocent. Surely that is not a proper question to pose to this witness. And was it not just a short time ago we agreed that Dobbs told your honor and myself he had no intention of hiding behind the Bible?"

Thompson looked at Charlie and then at Dobbs, "I am going to let this witness answer the question, if he wishes to. Asking a witness to declare a judgment of religion is a tricky slope to go down. I will allow it if he feels he can answer the question. But I would remind him and the prosecution they have a very narrow window of latitude in this matter. I would remind Mr. Dobbs he is very near that place with the Bible he said he would not go. Is that clear?"

Dobbs stated, "Yes, your honor. My intent was not to cross that line."

Thompson looked at Charlie, "For now we'll move forward. But be careful Mr. Dobbs. You may proceed with the question."

Dobbs nodded his acknowledgment of the judge's instructions. As he once again asked Gordon, "Do you believe Doctor Mathenson has broken the laws of your God or has she broken the laws of man?"

Gordon hesitated and looked around the courtroom. He paused as he looked at the delegation of religious men sitting in the back row. He could tell they were waiting for his answer with great anticipation. "The question you have asked me to decide requires more knowledge than I have at the present time. You, the prosecution, have laid out the guidelines for the charges concerning embezzlement, wire-fraud and misuse of government funds. The only way to judge those charges is to know what her work has discovered. As of yet no evidence has been entered for me to make such a judgement at this time."

Dobbs' face showed he had not anticipated that answer. He looked over a couple of papers in his hand and then said to the judge, "Your honor, at this time we have no further questions for this witness."

Judge Thompson looked over at the defense table. "Mr. Morgan, your cross of their witness."

Charlie sat for a moment, not moving. Then he looked at Downing and whispered, "We need a short recess." Downing shook his head that he agreed.

Charlie stood and said, "Your honor, we request a short recess before continuing at this time."

Thompson looked at the clock on the back wall, "We will reconvene in one hour." With that he recessed the court.

Charlie motioned for the others to follow him down the hallway to an empty conference room. As Jake entered he said, "Close the door, we need to talk."

In a moment everyone had given their attention to Charlie, "Well there it is, Clay has laid the cards on the table. He has said out loud what the jury has been thinking." He looked at Elizabeth. "He, along with the jury, want to hear from you what you have done." He then looked over at Angus, "I am sorry I asked you to travel a great distance thinking I wanted you to explain to the jury about her science. I figured if some members of the jury were not comfortable with the science, they may associate it more with you than against Elizabeth. But now, with Clay's answer, I think there is only one person who can answer the questions on everyone's lips. She told us last night about her science. I think it is time to let her explain it to the jury. Dobbs has concluded his witness list and it is our turn. I have made a mistake in thinking about this case as if I was still a prosecutor. I was ready to put Angus on the stand and spend a lengthy time establishing him as an expert witness. I may have taken a full day to do that."

He sat down across from Elizabeth, "I think our only play is for you to testify and tell the jury what you told us last night."

Angus, who was sitting beside Elizabeth said. "I have not been privy to the rest of the testimony in this case, but I think Charlie may be right. Those members of the jury have not stopped looking at you during Clay's testimony. I know from our conversations over the last six months you realized sooner or later this day would come. It is time. And I believe in my heart that is what you wanted."

Angus looked around the room and then at Charlie, "I can't help but think the prosecution has maneuvered this case to this point. From what you have told me of the testimony of the other witnesses and now Clay's testimony, he has moved this case from a monetary case to science. Dobbs has done it without making it outwardly apparent. We know he has played the science verses religion card. Only it is face down. He's gambling the jury knows he has an ace in the hole. He's a pretty smart cookie."

Elizabeth paused for a moment before responding, "I am not scared of the truth and what they think of me but of how they will judge the important work my research team has created. I am afraid that they will try and use religion to demonize the science. I don't know if I am the one to explain it to the world, but I know there is no one else. I am ready."

Charlie looked up at the clock, nearly forty-five minutes had passed since the recess began. He knew he would have to prepare his approach to questioning Elizabeth now that they were changing directions in her case. "We'll need a few hours to prepare. Jake, contact the bailiff to ask the judge for a recess until tomorrow."

Jake left the room as those remaining waited for an answer. It did not take long for him to return. "The judge agreed, but wants to reconvene the court to give instructions to all at the same time. That way he could assure everyone gets the same message."

Thompson reconvened the court and explained that more time had been requested by the defense. He then reaffirmed to the jury that they were not to watch, listen, read, talk or in other fashion seek further information about the case outside the courtroom. He explained that his instructions may be difficult to follow, with all the outside interest in the case. If they were to hear or see outside information in the press, it may be unsubstantiated and could even be purely speculations on the part of those reporting the news. The defendant's life, career and future depend on them judging her on the evidence given inside the courtroom. With those instructions he dismissed the jury until ten-thirty the next morning.

Chapter 26

Elizabeth's Truth

Just after midnight Charlie finished working to prepare Elizabeth for her testimony. He instructed her and the others to get some sleep, reminding them the day's court appearance was just a few hours away. Elizabeth needed her mind to be clear before going onto the witness stand. As Charlie was arranging his papers for tomorrow, he felt he needed some cool night air. In the mountains the sound of water and the coolness of the night helped him rest and clear his mind. He knew he needed more than just sleep before facing the jury. For some reason he remembered his own words from earlier that afternoon to the others, "I have made a mistake in thinking about this case as if I was still a prosecutor."

As he left the conference room Angus asked, "You going to be able to sleep?"

Charlie chuckled, "Probably not. I thought I would take a walk down by the river, unwind a bit in the night air, would you like to join me?"

Angus quickly responded, "I'd love too."

The yellowish glow from the streetlights gave the pavement a golden look in the night. Just ahead the river appeared to be a blackness devoid of light except for the reflection of the old suspension bridge's lights.

Charlie said, "I hope that I did not cause a problem by asking you to help out with Elizabeth's case. I was thinking like a prosecutor when I figured to put you on the stand to explain in great depth her research and its benefits for the world. But the jury wants to hear from her, especially after all the controversy surrounding this case. Hard to think they have been totally isolated from the stories and rumors. I think putting her on the stand is the only way out for her at this point. The only thing that may hurt her is the missing cryogenic canister Dobbs introduced. Everything else is explainable."

The two men stopped as they walked and looked out over the darkened river. Angus laid his hand on Charlie's shoulder, "You'll do fine. Just let her tell the story. She's a strong woman and only has good intentions. Her quest for understanding took her down a road I don't think she even envisioned in the beginning. As to the missing canister…have her tell them the truth. It's her story, let her tell it. Other than the one guy who just sits there with his arms folded, the rest of the jury is paying attention. Trust them to do the right thing. Her fate will be in their hands this time tomorrow night. Let's hope they understand her intentions."

Charlie pointed toward a bench and gestured for them to sit. After making themselves comfortable on the damp bench, Charlie asked. "Is this at all easy for you to understand when you look at it from your faiths standpoint? You are a scientist. A church vicar. That combination has to be a conflict within yourself, not to mention your religious beliefs?"

Angus looked at Charlie, "Easy for me to understand? I can't say any of us will truly understand it all. That's why we call it faith. Early on in my career, I was determined to be the first to say I wanted to know why we are here. I was an arrogant S.O.B. I went to Stonemore and got my doctoral in astrophysics. I studied the stars, looking for the answers. I was convinced they were out there. When I could not find the answers in the stars I went back to school and got my degree in biogenetics. I figured if the answers were not out there, perhaps they were inside each of us. Again I was frustrated with my work. You have to remember it was not too many years ago that understanding the human DNA was impossible without years of research. It was about that time I began drinking. I don't know why I thought I would find the answers in the bottom of a bottle, but I did. Then Tuck spent a month in a personal intervention into my life. I put him through hell. After a while in my anger and endless hang overs, I felt I began to feel some relief. Maybe the correct word was not relief, but I came to realize that I may not find the answers and that was okay. It still took another year for me to find my way back. During that year, I decided to go deeper into the ministry and see if Tuck was right. Learning has always come easy to me. So after a couple years, I was an astrophysicist, Doctor of Bio-genetic science and an ordained vicar. And I am a binge alcoholic. You experienced firsthand my latest relapse into the bottle when you visited me a short time ago. After all that, do I have the answers? I don't, but Elizabeth's discoveries have added more questions than it has answered. In the end, I rely on my faith to tell me it will be okay, it is part of His plan and I was not meant to understand it all. One thing's for sure, the answers are not in the bottom of a bottle, even if I do on occasion still go looking. "

Neither man spoke as they watched a barge pass silently in the near darkness. Finally after it's passing, Angus patted Charlie on the arm, "We had better get back to the hotel, ten-thirty will come soon enough. I don't want to fall asleep during the trial that might answer the greatest question mankind has ever asked about creation." He chuckled at his words.

Charlie looked over at Angus and said, "And what is that question?"

Angus stood and looked up at the stars. "Are we meant to find the answers that we believe only God has the right to know?"

Charlie stood and hesitated for a moment before he said, "If you are correct, isn't the pursuit of science the right to use our intelligence to challenge that question? After all, we are His creation in the first place. If we say Elizabeth's discovery is wrong, aren't we saying He made a mistake in giving her the knowledge to ask why?"

Angus smiled and said, "I am glad you are the one asking the questions tomorrow. Remember, her fate will be decided by the answers." The two men made their way back to the hotel for the night.

As Charlie and the others entered the courtroom, there was a sense of anticipation by those present, today is the day of answers. They knew the defense would begin their case for the jury. Charlie sat quietly looking over his questions as the courtroom filled with people. He could hear mumbling by a couple reporters hoping to get the scoop on today's testimony. He glanced up at the clock and showed it was nearly ten-thirty. Placing his hand on top of Elizabeth's hand he said, "Well, the press is here and in the back row are the religious scholars. All of them are waiting to hear what you have to say. Are you ready for this?" She smiled a little nervously as she shook her head yes.

At precisely ten-thirty the door opened and Judge Thompson rushed into the courtroom before the court bailiff could complete his announcement of his entry.

Thompson looked around the courtroom seeming to eye all that were present. "I would remind those of you in the audience that I will tolerate no expressions of opinions from the room during today's testimony, is that clear?"

Without waiting for an answer he looked toward the defense table. "Mr. Morgan, is the defense ready to proceed?"

Charlie stood and said, "Yes your honor, the defense calls to the stand, Doctor Elizabeth Mathenson."

Dobbs looked suddenly surprised at Charlie's announcement. He had anticipated that he would bring expert testimony from Angus Whitechurch. He had not anticipated that the defense's first witness would be Elizabeth. In the courtroom an audible murmur of surprised voices filled the courtroom. Thompson quickly slammed his gavel several times, admonishing the courtroom of his displeasure in the outburst of opinions. "Silence…silence. Any more sounds and I will clear the courtroom. Mr. Morgan, you may proceed."

Elizabeth made her way to the witness stand. She took the oath and sat down looking nervously at Charlie. Judge Thompson said to her, "Doctor, I wish to remind you that you do not have to testify in your own defense. Do you wish to continue?"

She looked at him and gave him a positive look as she said, "I understand that I need not testify. I am prepared to answer the charges made against me." Then she took on a defiant look as she looked toward the back row where Jedidiah Cain sat. After a few seconds of obviously staring at him she said, "All the charges against me."

Thompson shook his head satisfied and said, "Proceed."

Charlie approached the center of the area in front of the bench. "For the record, would you please state your full name?"

'Yes, Elizabeth Mary Mathenson."

"And can you tell us what your profession is?"

"I am by education a trained medical doctor. I am also a researcher in the field of bio-genetics."

"How long have you been a medical doctor?"

"I completed my residency in 1978. I immediately went into the field of research."

"Why did you not go into medical practice since completing your training?"

"I came to realize that practicing medicine was only placing Band-Aids on illnesses. I could see the system wanted doctors to be dispensers of pills to relieve the symptoms and the pain of the illness. There is a lot of money to be made dispensing drugs to patients. Some did nothing to prevent the diseases or afflictions ahead of time. I decided that if I went into research, I may be able to help develop cures for illnesses before they became incurable. The field of genetics was not new, but great strides were being made. I felt I could make a difference."

"Doctor Mathenson, would it be fair to say you are one of the top researchers in the world today in bio-genetic science?"

"I believe that I am among several top scientists who have advanced the future of the field. Genetic medicine is several different disciplines. Some are working on cloning of cells. Others are working with stem-cell therapy. The science of medicine to treat genetic abnormalities is yet another. To restrain the science to only one area of research is far too limiting for the science. Each day new applications of genetic medicine are being discovered."

Charlie then said. "Doctor Mathenson, you are on trial today, not for your research, but for alleged violations of the criminal code regarding embezzlement of federal grant funds, wire-tapping and misappropriations of federal funds." Charlie looked at the jury who was fixed on Elizabeth's testimony. "Can you answer the charges brought against you?"

Elizabeth hesitated for a moment, "It is true that I am here because of the charges as you stated, but in order for me to answer those charges it is impossible to separate my research from the question of the money."

Charlie looked at the jury and then back at Elizabeth. "Can you explain what you mean by your statement?"

Elizabeth then looked directly at the jury, making eye contact with each member as she began. "We have known for many years that particular cancers can run in families. Breast cancer and uterine cancers can be passed from one generation to the next. In men, testicular and prostate cancer can sometimes run in families. In some ethnic groups, cancer can be more prevalent than in the general population. Today, women are seeking to be tested for the defective BRCA gene in their DNA. If the defective gene is found, it raises the possibility they may develop cancer. Some women are choosing to have preventive surgery to remove their breasts and uterus before any possible cancer is found. This is a very devastating decision these women have chosen to make. The surgery will change their lives. But for these women it gives them a chance of being in control of their bodies. Their genetic heritage doesn't have to control them. But even with this drastic measure there is still the possibility they will pass the defective gene onto their daughters. Choices for men are not as easy when it comes to prevention of possible inherited cancers. But in this case we know what gene is the problem and science is making efforts to identify and correct that defective gene before it becomes a problem. I began my research looking for those answers."

Charlie glanced at the jury and then asked, "So, how was that original research funded?"

"When I first began, most of the funding came from small grants from drug companies. A couple national organizations for research into special cancers also provided some monies. The university genetics lab was well funded, primarily because it had made several discoveries in understanding how genes affect cancers. Then a few years after I began the research, my team made a discovery which got the attention of the federal government. We were able to find a marker in certain DNA strains that gave us our first real understanding of the evolution of a cancer in breast tissue. Luck would have it, about that same time the President made a commitment during his State of the Union speech declaring war on cancer. After his speech and our discovery, federal grant money seemed to pour in for a while."

Charlie interrupted her, "Now, at this point did you have any part in the wording of the grant request? That is to say, did you as a member of the team at that time help fashion the wording of what you intended to do with the requested funding? In a sense a mission statement?"

"No. You have to remember when this first began I was a member of the research group. Our work was governed by the board and the direction they wanted to go. In truth they monitored what we were doing and fashioned their grant money requests along the lines they believed our work could produce favorable results. Good result…more money. If they saw we were making strides for answers and having success, then they went out and sold our progress to encourage new donors to join our programs. For them, it was about the money coming in. More money, more equipment and more post-graduate programs for the best and brightest minds in medicine. It was a business for them. For those of us in the field it was still about the work."

Charlie emphasized his next question, "Now, to be clear, you had no part in those original grant requests. Absolutely no input at all?"

"As I said, at that point no. My only goal was the work. After that our research scope grew with each passing month. The influx of money gave the program great latitude to make improvements in equipment, people and even more important, more research money. After the second year our program was at the top of the game. Others from around the country and even the globe came and studied the work we were doing."

Charlie gave the jury a quick look as he turned to Elizabeth to ask his next question. "Ok, so the jury understands, the university's genetics lab grew with the input of funds and more professional people. The mission of the lab was simple, find answers to medical questions that could be answered from within the DNA of a man or woman?"

"Yes, I guess that is a simple way of putting it. But you have to remember it was not only our lab, but labs from all around the world were also on the quest to find secrets hidden within our genes. One other important event happened about that time, Scottish researchers announced they had successfully cloned a sheep using a somatic cell. A process of nucleus transfer. That event energized the science of genetics, but it also awakened those who already believed we were tampering with forbidden knowledge. Science saw it as a way to create life from a simple cell. Naysayers saw it as a way we were trying to build a sheep or a person with the traits we wanted them to have. You have to remember the horrible experimentation done to create a perfect Aryan race of superior people. Some saw the announcement of cloned sheep as the first step toward the goal that once cost millions their lives."

Charlie walked a few steps around the courtroom and then raised his hands, "Well, isn't that what you all were trying to do?"

Elizabeth responded sharply, "No! We were trying to understand the process of life. Study it so we could begin to understand how life begins as a single cell and grows into a complex living thing. The research to clone Dolly did something new. It removed the nucleus from a living cell and replaced it with DNA from another. It didn't always work, but it was the first step. This was ground breaking in genetic science."

Charlie allowed the jury to absorb her answer before asking, "Doctor Mathenson, at what point did you become head of the bio-genetics lab?"

"In 1999, the university reorganized the entire research department and separated out the genetic department. I had at that point become one of the genetics scientists that was well known in the world. I traveled and gave lectures on our research. At first the work in the lab was all hands on, but as we grew bigger there became a political side to our work. I never cared for that side of the lab program. I knew it was important to play the game in order to keep the money coming in for our research. So… as the head of the lab, every now and then I would be paraded out like I was part of a dog and pony show to speak about our work. Kiss up to drug companies CEOs with the hope they would invest more money. By 2012, we were considered to be the leading genetic research lab in the United States."

"You speak of the leading lab in the United States, are there others that are considered important in the scientific community for genetic research?"

Elizabeth thought for a second and then glanced toward those sitting behind the defense table. "Yes, I would have to say there are many labs doing important research work. There are a few programs we have worked with and shared important parts of our work. In England it would be Stonemore University under the direction of Doctor Angus Whitechurch. In Israel, it was the Institute for Genetic Research under the leadership of Doctor Abel Hasum. And in the Far East it is Doctor Lee Gyeong at the National Genetic Lab of Korea. Each of these labs are working on different aspects of DNA research."

Charlie picked up a paper and said, "In the complaint it sites as one of the indictments, that you directed funds into foreign research projects without any authorization. It alleges that you sought to move the money and then use it for unspecified reasons. Those projects where you directed funds were to a lab in Korea. Can you tell us why you would move the money without authorization into a program run by a foreign government lab? It also states the lab in question worked in association with entities in the Russian government."

Charlie looked up at Elizabeth. "Those are pretty strong chargers. Do you have an explanation why you did this?"

Elizabeth paused, gathering her thoughts. "Korea has for many years been working with the science of genetics. In their country the government does not limit the scope of their research, in fact it encourages it. They see the potential for the important use of science in the future. Their research had made some breakthroughs in advancements repairing damaged DNA sequences."

Charlie interrupted her, "Aren't you talking about their efforts to bring back to life a woolly mammoth?"

"Yes, but my interest was not in bringing back the mammoth. It was in the technology of repairing damaged DNA sequencing."

Charlie again asked, "What about their working with the Russian government, is that true?"

"Yes, I guess you could say it was true. It was more like they were working with the Russian scientist that found the mammoth's DNA in Siberia. They knew the Koreans were advanced in their research techniques, so they decided to work with them. Together they made ground breaking discoveries in the repair of damaged DNA sequences."

"Ok, let me understand this, you gave the Koreans funding to help with research to bring back a ten thousand year old mammoth, just so you could gain the technology to repair damaged DNA sequences? Sequencing you could use in the university's research to help repair damaged genes in hereditary cancers?"

"Yes, that is the general principle of what I intend to do with their science. I have discovered sometimes I can make faster progress in medicine when I don't concern myself with who discovered some aspects of the research. As long as it can be used to further our work. I invested in their project and in return I gained knowledge of a special protein enzyme. This molecule, when programed correctly, can repair damaged DNA matrix. Think of it as a 3-D bio-printer. It would have taken us another year to perfect this molecule to do the repair work. The Koreans gained funding and recognition from the United States. We gained advancements in our science. I can explain the process further if you would like?"

Charlie shook his head indicating, no, "Not at this time. A little later when you have explained more of the entire project. I believe then it may be helpful to the jurors. Now, once you had the knowledge of the molecule, did you at any time in the process sit down with the board or the controllers and explain your thought process in gaining this technology?"

"No, as I told you I am not good at being politically correct. To me it was just another dog and pony show. Besides, it was just another step toward the goal of the program for me. Why go before the board and tell them we are one step closer by using technology we learned from the Koreans? They would want to call a press conference to announce a breakthrough in our work. They were only interested in the big scientific announcement. With an announcement like that they would hope to draw more investors which meant more money. My team and I were interested in finding answers, not taking credit for something we did not do on our own. We just wanted to use it for our matrix."

"Ok Doctor Mathenson, you have explained your work with the Koreans. The indictment also speaks to money spent outside the project in Israel. Were they also trying to bring back a mammoth?"

Elizabeth got a bit of a frustrated look on her face as she started to answer. But before she could explain Dobbs stood and objected.

"Your honor, I fail to see where Mr. Morgan is going with this line of questioning. Dead mammoths? I believe the defense is trying to confuse the jury with farfetched science-fiction stories."

Thompson motioned for the two attorneys to approach the bench. "Mr. Morgan, I have an idea where you are going with this. My grandkids have me watching the science channel and I know what they are trying to do over there. But, don't you think you could leave this out and just have the doctor explain the points of her working with the Koreans that you believe are pertinent to her defense?"

Charlie leaned closer to the judge's position, "Your honor, I appreciate that you believe you may understand where I am going with this line of questioning. But it won't be completely evident until I conclude my questions for the doctor. While it's true her explanations are helpful for the jury to understand her innocence on the legal charges, I believe she also will be judged on information not on the legal record. Underlying charges against her are more than can be judged by this court. I believe in order for her to receive a fair trial, the ethical questions must also be answered at the same time as the legal charges. If you allow me to continue my questions, I am sure you will understand at the conclusion."

Charlie looked at Dobbs. "This is not the trial either of us expected when we were first asked to be part of it, but trust me to continue my questions in fairness to my client. In the end, all will be revealed and the jury can make a judgement on what I believe to be all the evidence in their minds."

Dobbs paused for a moment before answering, "Charlie, I have known you for a long time. I've never seen you try and pull the wool over a jury's eyes. I think you are going to convict your own client, but if you believe this is the only way to find answers in this case, the government will not stand in your way."

The judge motioned for the two men to return to their positions. Dobbs spoke, "Your honor, the government withdraws its objection."

Judge Thompson said, "Mr. Morgan, you may continue your line of questioning."

Charlie again addressed Doctor Mathenson, "Doctor, were the Israelis trying to bring back a mammoth?"

Elizabeth testily answered, "No! They were not trying anything such as that. My dealings with them has to do with a collection of DNA samples. Before 2000, scientists collecting DNA samples often used a variety of methods. Hair, fingernails, skin samples, blood, and urine. Science was still trying to find the best method for sampling to get a usable sample of DNA. Israel, like the rest of us, collected many samples wanting to study the genetics of their population. There are factions in the country that didn't believe science should be doing such studies. They believed they were using science and it was a violation of their religious beliefs. Not all scientists in the country felt this way. So, the research was often spoken of quietly, not to anger those on the religious fringes. As sampling methods improved, it was found that DNA could be collected from saliva or a swab from inside the mouth. They had collected many samples in the early days of their research and preserved them. It was decided they were no longer needed. But what to do with them? They felt they simply could not destroy them as if they were biological waste. A council was formed and they decided that they would offer to other scientific groups some of the samples. I got wind of this when I attended a conference in Tel Aviv. I was particularly interested in one group of samples they had in their inventory. They had collected nearly ten thousand samples from a religious sect in a part of Israel. They were of the Druze faith, a very honorable group of people. Their faith has elements of Judaism, Islam and Christian values. They believe in reincarnation after death. But, the unique thing about the faith is that it goes back over three thousand years. And for the last thousand years no one may join or leave the faith. I knew this special group gave me the opportunity to study a closed community's DNA that had been unaffected by groups of people outside their own community. It would be almost like a control group. I knew when I had the opportunity to purchase the samples I could not let them pass by. If I would have contacted the University for funding to purchase the samples and have them shipped, it may have been too late. There was the possibility some other researcher would discover the unique group of samples and purchase it for their own studies. So, I did the only thing I could, I used my credit card and made the purchase. I charged fifteen thousand dollars' worth of samples to my credit card. And yes, I did have to call the credit card company and explain so they would approve the transaction. Next, I shipped the samples to Stonemore. I had worked with Doctor Whitechurch and we were on the verge of a new way to analyze DNA samples using a new technique that would greatly speed up the process."

"So, you spent your own money on the samples and shipping to England. Did this credit card transaction pose a problem with the controllers back at the university when you submitted the invoice for reimbursement of the funds?"

"Yes, I would say it did in a couple ways. First, any expenditure above ten thousand dollars needed prior approval from the board. Second, such a large request for reimbursement to my personal funds was a major problem when the auditor came. And to make things even worse, the samples that I purchased were not in the university's lab. They were in England at Doctor Whitechurch's lab at that time."

Charlie then asked, "Was that the only problem with the samples?"

"No, shortly after I purchased them and shipped them out of the country a member of an extreme religious faction in the Israeli government found out. He was highly upset that a non-

Jewish person was in possession of the samples. He made accusations about experimentation on the DNA. At one point he even compared it to an act such as the Nazis experimentation on the Jewish people."

"Couldn't you have simply explained to them what you were intending to do with the samples? Or even better, returned them to the Israeli goverment?"

"By the time I knew about the controversy, it was too late to retrieve the samples in their original state. Doctor Whitechurch and his team had processed the samples. This was only possible using the new light emitting diode technique we had developed. Time needed was greatly reduced to analyze each sample. We offered to share the scientific data, but the political climate was so enraged. The Israeli government made it clear they wanted the original DNA returned to their country untouched by non-Israelis. There was no way I could comply by that point. I understood it was not the samples themselves as much as the political pressure to satisfy social factions within their country that were calling for their return."

Charlie took on a very serious look as he asked, "So… let me be clear on this point. The complete results of Doctor Whitechurch and your study of the DNA samples were offered back to the Israelis…but they refused your offer?'

"That is correct. Without any explanation as to why they refused our offer. There is one more issue that happened unexpectedly about the same time. When the shipment of the DNA samples was first received in England, Doctor Whitechurch contacted me and asked if there was more than one canister to be shipped. I explained I only knew of the one that contained all ten thousand saliva samples. I asked why he had made his inquiry as to how many containers were part of the shipment. He explained that along with the shipping manifest and address attached to the canister, there was a shipment tag that indicated it was one of two items in the shipment. We decided after discussing the tag that it must be from a previous shipment to another lab. It is not unusual to reuse special bio-containers when transferring samples from one lab to another. Then, a day later a second smaller container arrived. It was marked two of two. The second container apparently was held up for some reason. It was a special container with its contents preserved in dry ice. When it was opened they found additional DNA cell samples collected from the inside of the donor's mouth. This second container was a complete mystery. We knew it came from the same lab as the first shipment, but these samples were not part of the deal that I had made. It was clear, even with the controversies surrounding the first batch of samples, someone inside the Israeli lab wanted me to have these samples. I tried to make inquiries about the second container, but the lab in Israel insisted there was only one container and they wanted it returned untouched. A day later we received a manifest for the second container. The invoice showed an address that it had been shipped from. No such address could be found in Israel. We were at a loss to understand the second container. But, the documentation included bar code numbers that corresponded to the original saliva samples. To this day I am not sure why the second container was also shipped to the lab in England."

Charlie waited for a moment, thinking about Elizabeth's answer. Then he asked, "Can you speculate why someone wanted the second group of DNA samples out of Israel and into your hands?"

Elizabeth immediately answered, "No, I have no answers as to why. Perhaps they wanted the samples preserved outside of their country. In England the government and scientists have worked together to promote the science, but they have worked with the professionals in the field to come up with guidelines to help researchers work within the boundaries of the laws. They also have a long term-state of the art storage vault for preserving biological samples far into the future. Russia has also launched a project that can best be described as a genetic ark. DNA, biological samples and plant seeds are being preserved for the future. England is on a similar track, but not at the scale the Russians are building."

Charlie then asked, "The obvious question is, what happened with the second container of DNA samples?"

"We tried once again to contact their government department of science and they denied the existence of a second shipment. They continued to strongly protest that we had not returned the first and only shipment. With no luck in working with the Israelis, we decided to catalog the second group of samples along with the results of the first DNA analyses. After that they were placed in long term storage. DNA, if properly preserved, can be viable for at least a thousand years and maybe more. At the present time the samples are housed in that long-term storage site."

Charlie paused, then asked, "Before we move on, I would ask you this question. It seems there are several players in the scientific world that are interested in gathering DNA samples for long term storage. Do you have any idea why this would be taking place?"

Elizabeth hesitated for a moment. Then she looked toward the jury. "In Norway, at the Svalbard Global Seed Vault, approximately four hundred thousand samples are stored. Deep within the frozen land, the vault is home to a wide variety of seeds specimens we use to feed our world. The purpose of the vault? If for some catastrophic reason let's say a blight would decimate the world's corn crops, or a mold spore would infest wheat storage bins around the world, it would be possible to restart those crops with the seeds in this depository. We first would have to figure out how to safely reintroduce new crops so they too would not be affected by the disease or mold. You ask, why does there seem to be so much interest in these Arks in Europe? From a scientific standpoint you would have to believe they are preparing in the event crops or mammal DNA is needed at some point in the future. I don't believe they are predicting an urgent need. Europe has seen plagues and terrible famines devastate the European continent several times. There are those in the scientific community that believe if some unforeseen need would arise they are prepared. At one time oceans and vast mountain ranges prevented crop diseases and natural disasters from spreading from one region to another. Today distance over land or oceans is no longer a barrier to prevent global spread of a problem. Forward thinking governments in Europe are making some preparation in the event of a future catastrophic event."

Charlie asked, "Do they know something we don't? Maybe they are crying wolf just to make us think they know something."

"No, I don't think that is the case. Europe is made up of many countries that have been around a long time. The European history also shows they have experience blights and plagues over the centuries. Perhaps they are only being diligent in preventing future reoccurrence of mass diseases and crop blights. They look at things with an eye toward the future. Every scientist in this country worries about the possibility of a global pandemic which we all know now is more than possible. Not everything is about money, as it is often in our country. The Europeans are not predicting a major event or disaster, but if for some unseen reason something would happen, they will have been the wise ones."

"Okay, to the best of your knowledge, what countries are making preparations for storage of these types of samples?"

Elizabeth thought for a moment, "Well, let's see, of course England, Norway, Israel, Scotland, all of the Scandinavian countries to some degree. Russia has the largest depository at the present time. China is believed to have a large research and storage facility built into a mountain, but little is known of their program. South Korea and Japan are also cataloging and storing samples of DNA, but they are concentrating on studying DNA at the present time. Korea has made great strides in working with DNA. For a fee, they will clone your dog or cat. This service helps to fund their research which includes the mammoth project."

"Is there any effort to preserve in what I'll call a genetic ark project in our country?"

"No, not to the extent those countries which I mentioned are working to preserve specimens of DNA. Individual labs and universities are preserving some samples, mostly due to the research they are doing. The biggest data base of DNA information is collected within two separate sources. Commercial DNA services that will provide a person with their family history and possible health projections. The second is with law enforcement. It is not uncommon for those who are entered into the justice system to have their DNA collected and placed into a data system.

Charlie paused and then asked, "Is there an effort to collect DNA data to be stored for future research?"

Elizabeth, for the first time seemed to be uncomfortable with Charlie's question. "In our country privacy is considered a basic birthright. Random collection of the very essence of a person is very controversial. To collect, study and catalog DNA for any reason can be problematic. That is why when I had the opportunity to gain access to the Israeli DNA specimens, I jumped on it."

Charlie glanced at the jury. He wanted to see if they were still listening to what Elizabeth had to say. They seemed to be listening to her testimony with great interest.

Charlie then asked, "I have a couple more questions that need to be answered which are contained in the charges against you. First, the list of unexplained expenditures was for a trip to an area of Africa known as the Rift Valley. Can you explain why a genetic scientist would travel there?"

Elizabeth looked toward the jury, "Yes, whether you are a follower of the scientific origin of man or a believer in the biblical origin, this part of the world is where mankind is believed to have begun. I spent two months traveling from village to village collecting DNA samples from those who would give them. The people who live in these remote parts of the world are the direct descendants of generations of people who lived in the area long ago. Their heritage and DNA history extends back before written time. By studying their DNA, we can gain knowledge of their biological heritage, unspoiled from outside influences. Before I obtained the Israeli DNA samples, this was the closest I came to finding a closed society without major contamination of their DNA. It is believed by science that all mankind came from this region of the world. About seventy thousand years ago mankind ventured out of the region and onto the continent of Europe. Some traveled to the west, others went to the east. Over the centuries civilizations developed differently as the descendants of those first people spread across continents. All mankind originated in that area of the world. DNA strains remained close to their original patterns for thousands of years. But, mankind has long ago moved beyond the territory his ancestors once lived in for thousands of years. Today the entire world has become a melting pot of not only different races, but different DNA strains. Children born today can trace their DNA history back to points all over the globe. In just a few hundred years our DNA has been exposed to many other cultures. In the next few hundred years our DNA will undergo changes most people cannot imagine. For a few of us we believe we can begin to predict the future evolution of DNA."

Charlie could almost sense the tension growing within the courtroom as Elizabeth told her story. But, her last comment heightened the interest even more. Finally he asked, "Are you saying how research such as yours is going to affect mankind's future genetic makeup?"

Again Doctor Mathenson looked directly at the jury as if she were speaking to each person. "No, I am not referring to my science or others and how that may change genetics. And believe me, what someday will be possible is unbelievable. I am referring to the natural evolution of DNA."

Elizabeth paused for moment collecting her thoughts. Then she began once again. "A while back a fellow colleague at Stonemore England made a discovery while studying DNA. At first I was skeptical of his findings. But after our research into the DNA we obtained from Israel, I began to see merits in his conclusions. By taking those results and turning them into data, we simulated the evolution of the DNA's information based on the changes it has made over thousands of years. By reverse engineering DNA we looked back into our genetic history. With that information computers were then able to predict future changes to our DNA. Now, to be clear, I am saying we based our findings on what the computer predicts will happen in twenty thousand years if our DNA is left to continue on it present course. In earth's history that is less than a blink of an eye. For mankind it is many generations of people. The results were sobering to say the least. After those results were analyzed, Doctor Whitechurch invited a group of leading DNA scientists to Scotland to discuss his findings. Today, I along with a growing number of others are saying we need to fully understand nature's evolution of DNA before we go making changes on our own. Nature has a way, even within our cells, to have checks and balances. She manages to somehow know if a change in DNA takes place, how it will affect another part of the sequence. I even believe if she detects some flaw in that process, she will eliminate that damaged gene in most cases. Whatever the controlling force in the process of man is, it has for the most part done well. But, even Mother Nature can only repair so much. I believe she can work with man's blueprint and ensure we will survive far into the future. But she cannot predict changes from outside sources, including man's genetic engineering plans. She may see the changes we are making as damaged genes and try to repair them. Nature needs time, thousands of years to evolve our future genes. Man has begun making those changes. Over the next few decades he will accelerate his quest to improve mankind. I believe our future is balanced between what man does and what nature's evolution will do."

Elizabeth hesitated for a moment as she gathered her thoughts. "Forgive me… I'll try to explain our theory in another way. Let's say you were to take a printed page of text, placed it on a copier, and made a copy. Then you took that copy and made a copy of it. You repeat this process of making copies of copies over and over. In a short period of time you would begin to see the letters become less clear. The printed text would take on a slightly blurred appearance. And the more you repeated this, the more the quality would diminish. Now, let's imagine that when mankind first emerged over a hundred thousand years ago, they began with very sharply defined DNA patterns. Now, whether you believe in the Biblical account of how we began or how science says we started, we know either way we began from a small group of humans. Those first generations spread across the world. With each generation human DNA was reproduced many times over. Logic tells us those first civilizations were very similar in DNA strains. There is no way of knowing scientifically, but it is undeniable that the DNA slowly began to change and degrade with the passing of time. Now you have to remember we are talking about thirty million genes within a DNA structure. If only one or two of those genes are changed in each generation for one reason or another, the next generation will carry that altered gene. Those isolated changes were confined to that group of people and their society. Another event in man's history that contributed to more changes in DNA was the practice of inbreeding that took place in a few early civilizations. We know that inbreeding was probably not uncommon in isolated cultures. Especially in what we would refer to as the upper class. What preserved records of ancient history we have, tells of how royal bloodlines were preserved within families. How that class of people chose to reproduce was the first attempt at controlling DNA. They believe they were of divinely pure blood. To have offspring from a lesser class of people would taint their bloodlines and displease their gods. With the passing of time, society grew and progressed; those practices have faded. In fact, we look upon them today as a taboo in civilized society. So, closed family bloodlines faded, but the fact that it once happened played a part in altering DNA. That altered DNA continued into future generations. As I said earlier, about five or six hundred years ago man began spreading across not only the lands he was born into, but he began seeking new worlds across far oceans, and tall mountain ranges. As cultures met they began to join together once again. Now, you have to remember that even though thousands of years had passed and countless generations had come and gone, DNA patterns in each of those people can be tracked back to their ancestors where it all began. With each generation, genes contained within DNA changed ever so slightly. That most ancient of blueprints within each of us for our own creation is very much part of our DNA today. When we look at DNA we see a complex chemical molecule changed by the environment of time. It also has been changed by the mixing of different DNA sequences. Some of those changes Mother Nature created on her own. In the complex molecule we find genes that somewhere in time were changed or became dormant. A gene once programed to support our immune system can mutate and change its function. Today it may help to create mutated cells that can grow into cancers or other abnormalities. Do you remember I mentioned the BRCA gene earlier? There are two of these genes and their role is to find cells whose DNA is damaged and make repairs. Parts of our DNA were programed to repair damaged cells in the beginning. What if over time those beneficial genes, themselves become mutated? Then they could no longer perform their job. This could increase the risk of cancers. It is certain these genes have been part of the body's repair system since the beginning. But sometime in the past one person possessed that gene mutation. We know that our body's cells can be damaged by external environmental events. It is possible that a man or woman two thousand years ago may have labored in the field under the intense rays of the Middle East sun. They may have been exposed to some chemical elements in their water or food supply. These outside influences could have made their delicate DNA susceptible to damage. Within a single cell a particular chromosome may have been changed. Those peoples' constant exposure may have given them some immunity and they survived. But the damaged gene became part of their biological make up. It passed on to the next generation. Along the line it mutated even more. Eventually, it became the gene that carries the mutation we see today. We believe the gene could not have been part of the normal progression of that particularly altered gene. If it had, it may have been carried in a man or woman's DNA around the earth. Without modern medicine, mankind would have died out long ago with this one simple gene mutation. Let's say if the BRCA gene had been in every woman, for thousands of years, it is likely female populations would have had a very difficult time surviving. Without the survival of woman, man would surely be lost also. Everything we are is contained within those twenty three chromosomes and millions of genes. Our very existence is contained in our DNA blueprint written in a complex mixture of chemical molecules. It is a very old recipe for the creation of mankind. We hope by creating genetic arks around the world we can preserve that recipe for the future."

Elizabeth paused for a moment and looked at the judge, "Your honor, may I have a glass of water before I continue?"

Thompson, who had been drawn into her explanation, hesitated for a moment before saying, "I think we all can use a short break. Let's be back in our seats at one-thirty." With that he tapped his gavel and disappeared into the back.

Charlie waited for Elizabeth to come down from the witness stand. As she came close he smiled, "You are doing well. I believe at least you have the jury's attention. Not sure how they are going to respond. You told them a lot and I am not sure they can grasp all the science behind your explanation."

Elizabeth smiled and said, "Thanks, I need a glass of water and a moment to think before we begin again." With that she returned to the defense table.

Charlie paused for a moment to think of his next move. Then he heard a familiar voice coming from the audience area. "I suppose next she will be telling them about dinosaurs and how some day they really will be in a park somewhere. Your client is acting like the world is run by science. She better remember that

God and only God knows our final destination and she and none of her fellow scientists can change that. Do you hear me CHARLIE MORGAN?"

About then Dobbs stepped up to Charlie's side, "Reverend Cain, I would advise you to step away and return to your area of the courtroom. If you don't, I will call security and have you removed. I think you were about to make a threat toward the defendant. She may be on trial, but I won't permit that kind of talk in this courtroom. Do you understand?"

Cain stared at both men before turning and walking away. Dobbs said, "Charlie, let's hope this doesn't get out of control before we finish this case. To be honest, I think your client is going down a path that will get her in trouble with both science and religion. And you seem to be helping her. I'll not stop her testimony, she seems to know where she is going with this. Whatever those jury members decide, it is going to be on her shoulders."

Charlie responded, "What is it they say, 'Sometimes the truth will set you free?' I think she wants to have it all out there for everyone to judge. We'll see what the jury thinks of her revelations of science and religion."

Time quickly passed and the courtroom quieted down once again as the jury was led back in. A moment later Thompson hurried once again into the courtroom. "All right, are we ready to resume the defendant's testimony?"

Charlie looked at the judge, "Yes, your honor, she is."

Thompson adjusted himself into his chair as he said, "Good, let's get about her testimony. You may continue."

Charlie stood looking at the jury with no expression. Then slowly he turned to Elizabeth. "All right Doctor Mathenson, you have explained to the court a great many things today. Your explanation as to the charges against you may be a bit murky for the jury. Can you put all the pieces together so they may understand your intentions?"

Elizabeth paused and closed her eyes as she thought about Charlie's question. Then she looked at him and then the jury. "I want to explain about a discovery I made while doing my research into cancers. I needed a protein from a bacterium to isolate damaged pieces of gene. I came up with an engineered molecule I called A.D.A.M., Short for Amino Dimatrix Atom Molecule. There were other proteins that science was using to do similar work. When I was in Korea I took part in repairing mammoth genes. It was then I began to understand the process needed to create a structure the DNA could use for its own repairs. Korean scientists were splicing modern elephant DNA into the damaged sections of the mammoths. But my theory was not to replace but insert 3-D scaffolding matrix, then the DNA itself would interpret its needs and make the necessary repairs. In essence, give the natural process a little help. We would not need to use genes from a living mammal, but the natural process within the genes would evaluate repairs they needed. In a sense, a good gene would be encouraged to reboot damaged genes back to their original state. The DNA could make its repairs, even when badly damaged by disease or time, as in the case of the mammoth. It was my hope that by naturally rebooting the DNA from its own blueprint, it would not set up a conflict within the DNA, as I believed genetically engineered genes could. The research using the ADAM concept eventually worked. The DNA's had its own ability to understand what repairs it had to make. It was also about that time stem-cell research was taking off. Unfortunately it was surrounded by controversy; was it a violation of life? When the sheep was cloned from an adult cell, it took away some of the problems. This discovery helped to reduce the disagreement within the religious community about stem-cell usage."

Charlie asked, "Can you briefly explain what a stem-cell is?"

Elizebeth turned toward the jury. "Each of us are born with an abundance of stem cells. Each awaiting instructions from the body. A stem cell is a blank cell that can become what your body may need. A skin cell, brain cell or a new muscle cell for your heart. As you grow older you have less and less of these cells to make repairs within your body. I knew the procedure to change an adult cell into a stem cell, but I began to wonder how the cell itself knows the information of what to change into. I came to the conclusion that in a stem cell there has to be a gene that contains every bit of information needed to create a person. In a sense, a master blueprint for our DNA. Something far more complex that even the DNA sequence. When we realized what we were looking for, we set about studying the Israeli samples. Next, a comparison to samples from the Rift Valley area of Africa, along with several other isolated cultures in our data base. Finally, and only with the use of massive computers at Stonemore, we discovered a single isolated gene that is similar in all those samples. Using an electron microscope we studied the gene's structure. We called it, AG1. The Adam Gene. It is the key to life as we know it. With this gene and our ability to create a 3-D protein matrix, our research team devised a way to repair damaged DNA sequences. We realized we may have found a way to repair mutated cancerous cells. Our discovery could very well be the key to all genetic disease."

Charlie stopped her there, "Well, that is a great thing. What you're telling me is, that you did just what the university wanted your lab to do."

Elizabeth hesitated. Her hands began to sweat as she now looked a bit troubled, glancing at the jury. "I felt great joy at what our research was on track to accomplish. The board of directors got wind we were on the verge of a great discovery. They encouraged me to publish a paper, announcing to the world our accomplishment. Late one evening while going over the results I began to wonder, what if I used the same technology to create an entire synthetic DNA matrix. I could use the original coded information from the AG1 gene that we had identified. I was sure that the original code was undamaged by time and conditions. It was the original blueprint for all mankind. It had to be. It was found in each of the samples we researched. Before each DNA is created it gets its basic information from this gene."

Charlie looked at the jury with a bit of a confused look. "Well, if that is true and we all came from one master gene, than each of us would be identical, correct?"

Elizabeth sat forward on her chair, almost eager to tell her story now, "No, that's where the beauty of chemistry of DNA comes into play. We know the complex chemicals used in the DNA's make up. But, it is like when constructing a building. It uses steel, concrete, wood in its basic construction. By using the same component, different builders can turn the same basic materials into multiple designs. It is the same with the original gene blueprint. When assembling begins, DNA needs four basic chemicals, adenine, guanine, cytosine, and thymine. These chemical components are referred to as the letters A,G,C,T when we speak of them. Your DNA contains six billion combinations of the letters. You may have seen a DNA's double helix formation. It is how your DNA plays its part in your body's make up. The same chemicals make up all our DNAs but how they are arranged is what makes us different. Little bit of guanine and you have red hair. A little less of thymine and you are tall. The whole process is complex."

Again she paused, gathering her thoughts, "Okay, I was telling you about our original discovery using the 3-D scaffolding matrix for repairs to mutated genes. I then went beyond the goals of that project. I wanted to see if I could use both the discovery of the AG1 gene and the A.D.A.M. protein matrix to create a complete synthetic DNA sequence. If I could, then I would insert it into a living cell."

Charlie interrupted her, "Now, you are talking about the cryogenic canister you purchased that contained human eggs. Is that correct?"

"Yes, that's correct. I saw on a medical auction web site a cryogenic storage container for sale from a bankrupt storage company. It stated the container was still connected to nitrogen cylinders. I recognized that type of cylinder as being the type often used in fertility clinics. I took a chance and placed a bid I knew would buy the canister. I was successful with my bid. I then gave instructions that the canister was to be shipped connected to the nitrogen and by expedited transport. When it arrived it had all the proper documentation from the bank auction. It also included the documentation it had been discarded from a fertility clinic. The specimens inside were no longer wanted by their owners. It was the storage facility's responsibility to make arrangements for proper disposal of the contents. For some reason that did not happen and the canister, along with its contents, were sold off when the facility went bankrupt. So you see, I came by those one-hundred and twenty five specimens properly. Even though they were to be destroyed by the auction company."

Charlie glanced toward the jury, He could tell they were not comfortable with that answer. He had to at least make it clear she understood what she was doing when she kept the specimens.

"Doctor, then it is correct to say that even though the canister ownership came to you properly, you did realize that after reading the documentation the contents were to be properly destroyed. They were for one reason or another, human eggs that were no longer viable or they were not wanted by their owners. Is that correct?"

Elizabeth hesitated for a moment. "Yes, that is correct. But you have to realize I saw an opportunity to further my research. For me to go out on the market and buy that many specimens would take weeks if not months. And to get proper authorization to buy the eggs could take a couple years in court. By a stroke of luck, and not by my doing, I had what I needed to continue my work."

Charlie quickly asked, "Continue your work? Did you see this as an extension of your research into genetic cancers?"

Elizabeth thought for a moment as she formulated her answer. "I would have to say yes, but understand that when you work in the field of medical research, sometimes answers come in a different form than you had anticipated. Using the AG1 gene placed within our created matrix could be a way to grow new organs from a single cell. My research was a step up from that basic idea. It could be the answer to a couple problems. I know it looks like I was out on my own in this work, but I truly wanted all the answers. If anything, I became too ambitious trying to rush and find those answers."

Charlie looked toward Clay sitting near the front row. "You know who Clay Gordon is?"

Elizabeth looked at Clay as she answered, "Yes, I have known Clay for many years. He is a professor of law at the university."

"Did you ask him last summer at a meeting if he knew a way past the university getting control of a patent process?"

She hesitated as she looked down at her hands before answering Charlie's question, "Yes, I asked if he knew any way around the process. I did not give him specifics of my work, only that it involved a new genetic process and it had ethical questions that needed answered before the university took control of the work."

"And what did Mr. Gordon tell you?"

"He was very clear. The contract we work under clearly states that any discoveries or patents will become the property of the university. Only research and discoveries funded specially for the government under the national security guidelines could be excluded."

"What did that mean to you?"

"I decided then that if my research worked, I would not allow my AG1 gene discovery and the matrix to become the university's property. I convinced myself that I could withhold the technology and still reveal the research that helped to prevent mutations in cancer cells. I was convinced I could do both."

"And did you?"

"Yes, the procedures worked and I created enough synthetic DNA sequences to place one in each of the eggs that I had obtained from the cryogenic canister."

"After you completed your work, what did you do next?"

"I moved the canister and documentation from the lab, to England as part of a shipment of tissue samples. There they were placed in long term storage in hopes someday the embryos in the eggs may be a new beginning if the human race would need them."

"Correct me if I am wrong. All that you have told us is untested at this point. In theory your new synthetized DNA cells may not work at all. Isn't that true?"

Elizabeth stared straight forward as she said, "Yes, those specimens sent to the ark for the future are in theory viable embryos. But, will they come to term in some future generation? I can't be sure. I do know they passed every developmental goal we set in the lab."

Charlie looked at the jury, he could see in their eyes they were in deep consideration of all she had explained during her testimony. He then wanted her to answer one more question.

"Doctor Mathenson, I have to ask, you could have walked away at any time in your work. You, by your own admission, chose to continue your work at the same time you were venturing down the path of unknown answers, even now. Why?"

Elizabeth looked at the jury. "When I began my work I wanted only to help save people from suffering with horrible diseases. And I believe I have. I have spent my life inside a lab with only one goal in mind, finding answers to some of life's biggest questions when it comes to health. As I drew closer to those answers, I came to realize the world outside my lab was only interested in money and who controls medicine. Would those people who could not afford the benefits of this research have access to it? Or would only the very rich benefit? Remember those ancient royal families that kept their bloodlines pure by only breeding within their own families? Could my work in some ways be compared to that kind of thinking at some future date? And what of Doctor Whitechurch's predictions, that in twenty thousand years, our own degenerating DNA will bring an end to mankind? At that point I realized the ark projects around the world may be the only salvation our humanity has. Perhaps those synthetic DNA eggs now in cryogenic freeze may someday be a kick start for a new generation of mankind. I don't have the answers. At least I have given those who come along in the distant future options. Someone asked me a short time ago, "Do I believe in God?" I have to say I do. And if I do, I have to believe that He gave me the intelligence to do the work I have done. Why else would He have placed me on this earth? I don't believe he would fault me for using science to do this work. When man first started asking for answers the first question he asked was, "Can I live forever? Then, a man two thousand years ago stood before them and told them, if they believed in him there is life in the ever after. Who is to say that the steps I have taken are not the steps into eternity? I will confess to all of you here that I am the one who anonymously started this whole thing. I gave a tip to the justice department that there were questions about my department's finances. I uploaded bits of information to online bloggers that continued the story. And yes I even tipped Jedidiah Cain to the inside information that has led to the disruption outside this courthouse. I have even withheld information from my own defense team. And for that I am sorry, but I wanted the world to know what I have done and what is being done by others. If what we do is not done with an eye toward the ethics of mankind and God, then where will our future take us." Elizabeth stared at the jury and then looked around the room. Silence filled the air as no one could believe what they had just heard. Finally Charlie broke the silence.

"Doctor Mathenson, I believe there is one more question we all need to know, since you are revealing all. Have you created new life?"

Elizabeth paused for a moment before she responded, "That question is for the future. In that cryogenic container in England's Ark Project, there are human eggs, one hundred and twenty four of them. Someday, someone will try and implant each of those eggs into a human. Will my research create life? I hope so. What will those children be like created from synthetic DNA? I don't know. But they will be the future. Those children, along with others that will now be created around the world, will be born in a world where this question has long been answered. I hope my actions will begin that conversation."

Charlie stood looking at Elizabeth and then toward the judge, "Your honor, the defense rest its case."

Judge Thompson hesitated as he looked at the prosecutor's table. "Mr. Dobbs, do you wish to cross examine the defendant?"

Dobbs was slow to stand. Finally he said, "Your honor, I am not sure even where to begin to cross examine Doctor Mathenson. I believe she has answered all the questions I have. And the ones I should ask, I am not sure she or anyone else can answer them. I would ask that we recess for the evening and return in the morning for any cross that I would have. Doctor Mathenson's testimony is so far reaching, I need to review my notes before proceeding. Your honor, may we approach the bench?"

Thompson motion for the two attorneys to come close.

Dobbs began, "Your honor, if you could recess until in the morning I believe we can wrap this up tomorrow. I'm not sure how I would question the doctor any further about her testimony. I have never in all my years been confronted with this situation. I will tell you that I am inclined to let her testimony stand and go into closing statements. Charlie, would that be a problem for you?"

Charlie responded, "No, we will be ready for our summation. I think it's all on the records. Not much more can be added."

Thompson moved his lower jaw back and forth for a few seconds, deciding what to do next. "Alright gentlemen, I will recess court until nine thirty in the morning. That way if Mr. Dobbs decides he does not want to question the witness, we can proceed to closing statements. I would like to have this in the hands of the jury by lunch. I trust neither of you intend a long closing?" Both men indicated they did not.

Thompson then directly asked, "Are we in agreement?" Both men answered in the affirmative. He motioned for them to step away from the bench.

Judge Thompson looked out over the court and then toward the jury. "I am going to grant a recess until nine-thirty in the morning. I don't believe I have to instruct the jury as to their responsibility in not seeking outside information about this case." With that he tapped his gavel and the court adjourned for the day.

Tucker and Angus walked to where Elizabeth was coming down from the witness stand. Without a word they joined hands and lowered their heads in a moment of prayer. Charlie believed she would need all the help she could get after her testimony today. She glanced at Charlie as the two others prayed. Elizebeth knew that he understood she was the information leak since the very beginning of the case.

Chapter 27
The Summations

Morning seemed to come early and no one had much to say as they entered the courtroom. Charlie had spent most of the night preparing his summation to the jury. He realized somewhere in the late evening he would be giving a defensive summation for the first time in his career. And even more challenging, it was a summation for a case no one had ever been part of before.

Downing sat down next to Charlie and asked, "What do you think the jury thought of her testimony?"

Charlie finished making a couple notes before he answered. "Well one thing's is for sure, when she told them she started this to bring her science into the light, it surprised everyone, including me. I know she was playing the system to get her research out into the world, but it would have been nice to know before everyone else did."

He hesitated for a moment as he thought, "But, as to how the jury took her testimony, I think they took it in stride after a few minutes. I watched when they were let go for the day. They didn't seem to be in a big hurry to talk with each other. That's a good thing. As to the rest of her testimony, if their faces were any indication of what they were thinking, I believe she kept them interested in her story. I saw no real disbelief in their faces. The guy with the folded arms and the one lady near the end seem to be having trouble keeping up with the facts as she spoke. Jake watched the faces of the jury yesterday and gave me a summary of what he believes they were feeling during her testimony. So far they have found her to be credible, at least in their expressions. He only detected one area they found a problem with her actions. When she tried to explain the cryogenic canister. It was evident from their expressions they found no fault in how she came to purchase the canister. But, when she continued working with the frozen eggs, they may not have liked that. In fact, one woman seemed almost disgusted by what she did. I am guessing that lady may have some deep religious ideals. Knowing this, I have to try in my summation to convince her and the others that Elizabeth's intent was for science. Her decision to insert synthetic DNA into the eggs was to find genetic answers related to her research."

Downing asked, "But, wasn't that what she was doing when it comes right down to it? Maybe not with a God syndrome, but she was thinking her scientific discovery could be part of the future of man?"

Charlie responds, "I don't think so. And I have to in some way make that clear to the jury. She predicted that a possible evolutionary path for our DNA may take it toward a time when it may not be viable to sustain mankind. She also said that could be thousands of years in the future. She wasn't crying wolf about something happening next year or even in our lifetime. She was saying we still have a chance to fix this, but it will require both science and nature to better understand how each works. The embryos in that canister are only one possibility, an insurance policy. Besides, it is untested technology for creating synthetic DNA. Can you imagine if she had someway figured out how to bring one of those embryos full term? No, I don't even think she believes it is yet possible. If events surrounding this whole affair are any indications, we as society have some soul searching before we are ready to acknowledge that it is possible. Those one-hundred and twenty five embryos represent an example of what is possible."

About then, through the side door, Samantha and Jud escorted Elizabeth to the defense table. Charlie tipped his head silently thanking the marshals for their help during this case.

Samantha leaned toward Charlie and in a whispered voice said, "They stocked the Smoke Hole Canyon yesterday, two truckloads of prime trout. Some problem at the hatchery and they wanted to move the fish quickly."

Charlie looked up at her with a disgusted look as he said, "Gee, thanks Sam. If my day was not going bad enough you just added to my frustrations." He grinned at her. She smiled back.

The courtroom seemed even quieter than it had over the past few days. Those in the room had an eagerness to hear what the two counsels have to say in their closings. As Elizabeth sat down, Charlie saw a nervousness in her face. He laid his hand on her arm in an effort to reassure her she was not alone. "Well, it's up to me to convince the jury you are innocent of the federal charges. I hope that I can also help them to understand their judgment should be about the money, not your science. I wish you had confided in me earlier about what you were up to. I am not sure if it would have made any difference, but I didn't like wondering who was leaking information."

Elizabeth answered, "I am sorry. I did not know you. Remember you walked in off the street wanting to help. I did not know if I could trust you. You made a reference to the summer of 81. I figured Bill had a hand in your coming to help. But under no circumstances did I want to involve him. I knew when I moved that canister from the lab and shipped them to England my career was most likely over. But I still would do the same thing. The world has to know what is taking place in research labs."

She hesitated and then asked, "Do you think Dobbs will put me back on the stand?"

Charlie thought for a minute before answering, "I don't think so. This case could still blow up in his face. Remember he has higher goals and being labeled the attorney who replaced God with science…that would not be a red letter item on his resume. No, I think he will let the chips fall where they might. For now, the burden of guilt will be decided by the jury, not some courtroom antics by an ambitious prosecutor."

About then the court bailiff announced, "All rise the Honorable Judge Thompson's Federal Court is now in session." As usual Thompson seemed to burst into the courtroom and into his seat.

As he sat down, he paused and folded his hands together, looking at the crowded courtroom. He then looked at Dobbs. "Has the prosecution made a decision whether or not it wishes to question the defendant at this time?"

Dobbs stood, "Your honor, I have reviewed her testimony. I believe the unusualness of this case has already given the members of the jury plenty to think about in their deliberations. The prosecution rests and is prepared with its closing statement."

Judge Thompson then looked at defense team. "Mr. Downing, is there anything more to be done at this time before we move on to the closings?"

Charlie nodded to Downing to answer the judge's question, "No, your honor. We are prepared to move ahead with our closing statement."

The judge made a few notes and then sat back into his chair, finding a comfortable spot. "Mr. Dobbs, you may proceed."

Dobbs stood and adjusted his coat. He then stepped around the table and walked toward the jury box. "Ladies and gentleman, when you were seated for this trial you were told it was to sit in judgment of a federal embezzlement case involving Doctor Elizabeth Mathenson, the defendant in this case. Each of you assumed it would be a straightforward case involving money. The government has presented to you information concerning how grant programs work. How they are administered and how they are to be utilized in the programs that receive them. I would have to say after hearing the explanation of the experts in the field of grant money, you should have a clear understanding of how they work. Next, you heard from the university's controller. She explained how the monies received from grants is distributed and controlled within the research projects. And especially how the money worked in the genetic lab's programs. You heard from the controller how Doctor Mathenson was difficult to work with and had little regard for proper protocol in the use of the money. In fact, she made the control of the money for the project downright near impossible. Next, you heard from Clay Gordon, longtime law professor at the university. He explained how Doctor Mathenson came to him and asked a hypothetical question about ownership of a possible discovery and how it could be hidden from the university board of directors, hypothetically speaking. Then in one of the most unusual testimonies I have ever witnessed in a courtroom, Doctor Mathenson, herself testified as to the case against her. She willingly admitted that she used federal funds to support foreign research that was hoping to bring back a woolly mammoth. And let's not forget this work was in coordination with agents of the Russian government. She funneled funds from her own project toward that research on two separate occasions. She told you that her only objective was to gain knowledge to benefit her own project in finding a cure for cancer. If that was her only goal, then why didn't she go through proper channels to use the money? Do it the right way. If she had...if she had I don't believe she would find herself in the situation she is facing today. Next she wants you to believe she went to Israel to a scientific conference and just happened to run across a yard sale of DNA. I can just see it now when she made the deal with the lab to purchase the samples. It must have been like when you go in to buy a car and the salesman says, "What will it take to put you in the car today?" Charlie noticed several of the jurors were shaking their heads; they had experienced such a car salesman.

Dobbs paused, looking at the jurors as he rubbed his lower jaw as if he were thinking, "Can't you just hear that Israeli lab personnel saying to her, 'Doctor Mathenson, if you take these specimens today for the low price of just fifteen thousand dollars, we'll throw in free shipping and handling? And if you act within the next twenty minutes we'll throw in a surprise gift you will receive later. How about it doc, we got a deal?' Lucky Doctor Mathenson travels with American Express. She got her deal. It doesn't matter that she knew her expenditures were above her spending limit without prior approval. She simply used her credit card and made the purchase on her expense report. Then…then she did not ship them back to her own lab, no, she shipped them to a lab in England. Supposedly they had some new technology to quickly analyze the samples. Finally she went out on the internet and found a deal on a frozen canister of human eggs that no one wanted or even knew still existed. Everyone but her believed they had been properly disposed of. She took the risk of buying the container sight unseen. Lucky for her the eggs were still intact. All of that is about spending money outside the confines of the research program she directed. Then you heard in her own words that she decided to save the world. Using her technology, she created one-hundred and twenty-four embryos for the future of mankind. What a wonderful thing she must believe she has done."

Dobbs movements had become exaggerated during the last few minutes of speaking. He had taken on the airs of an evangelist preacher, saving souls at a tent revival.

Dobbs stepped closer to the jury as he turned and pointed his finger directly at Elizabeth, "Remember… she was supposed to find a cure for cancer, not save the world. At what point did she appoint herself to be the salvation of mankind? No, not just the salvation, but the creator of new human life. Somewhere in England are one-hundred and twenty four new lives frozen in time. Were they created by God? No."

He pointed accusingly toward Elizabeth. "She created them!"

Dobbs walked back to his table and picked up a folded newspaper. He opened it and looked at the front page.

He began to read. "Modern Day Monkey Trial. That's the headline in the paper. It goes on to say a doctor has found the answers to prolonging life by a new scientific process. The question of eternal life has been found. Found not in man's beliefs, but in the hands of science. It is also said that she has for the first time in man's history created new life within one-hundred and twenty four embryos."

Dobbs turned to the jury, "I examined the complete story, not one word about she has found a cure for cancer."

Dobbs slowly walked back to his table and laid down the paper. He then turned to the jury. In a softer, more subdued voice said, "I am sure Doctor Mathenson set out with noble intentions. She is undoubtedly one of the best in her field. But she cheated the system and used funding that was to go toward finding a cure for cancer. Instead she used it for her own quest to find a type of immortality for her created children. Those are the embryos she has placed in a cold, frozen, stainless steel canister… well, let's just say it…she has created one-hundred and twenty-four Adam and Eves. Created not by God, but by her own hands. She not only played the part of financier for scientific research, she played the part of God also."

In the courtroom you could hear the muffled sounds of voices whispering not believing what they had just heard. Judge Thompson struck his gavel hard several times. "Silence…I said silence! Any more outburst of opinions whether out loud or whispered… I will clear this courtroom. Is that absolutely CLEAR?"

The room quieted. Thompson, not hearing anyone willing to challenge his authority, said, "Mr. Dobbs, do you have anything more to add?"

Dobbs stood for a moment as he looked at Elizabeth, "Remember you are sitting in judgment of Doctor Mathenson. You cannot separate the questions of her ethics and money when making your judgment. And you don't need to feel any remorse about having to make that judgment. Remember... she made that same judgment and she chose to forget the ethics of her obligations to the university and her profession. And even her obligation to mankind." With that he slowly returned to his seat and sat down. The courtroom was dead silent.

Charlie leaned toward Downing and whispered, "I was wrong, Dobbs just made the biggest play of his life toward becoming the next conservative Supreme Court justice."

Thompson sat for a moment looking at Charlie, "Mr. Morgan, are you going to represent your client in the closing?"

Charlie stood and said, "Yes, your honor, we are ready to proceed."

Charlie walked toward the men and women of the jury. He stopped about six feet in front of the box. He clearly did not want to crowd them after Dobbs' dramatic closing. He smiled at the jury as he began. "If you are like me, you may have had a difficult time sleeping last night. The anticipation of today's session weighed heavily on my thoughts. It may have been on your minds also." He moved forward and patted the rail that ran in front of the jury box with a reassuring gentleness.

"Sitting in judgment is not an easy task. I'll bet some of you may not have been eager to serve on this jury, knowing this day would come. The day you have to decide the fate of another human being." He gave them a reassuring look, hoping to ease their tension.

"Last night as I sat pondering what I might say this morning, I thought about each of you and your understanding of Doctor Mathenson's case. How could I help you understand she was not on a quest for personal greatness? But, on a lifelong mission to help save lives. I decided I would tell you a little about her life. Early on while she was still in medical training, she realized lives often could not be saved. The only answer was medication to ease the pain and suffering. Death was the only conclusion. Cancer was often too far along by the time it was detected. She realized there had to be better answers. She knew from her training that one single cell mutated somewhere along the line. It not only mutated, it survived. Just one damaged cell. She believed there must be a way to find that cell and stop the disease. It was at that point in her career she chose a different path after medical school. She wanted to try to find those answers. For her entire career she has worked in that lab at the university, searching for answers. The prosecution wants you to believe that Doctor Mathenson obtaining established technology from Korea was a bad thing. Mr. Dobbs wants you to think she brought that knowledge back for her own selfish purposes. The truth is, that proven technology hastened the pace of her department's research into understanding the DNA matrix. She used good managing skills. As director of the lab, she is expected to make those calls."

He paused and looked at the big man with his arms folded in the back row. "If you were to have a new house built, would you build the kitchen cabinets yourself? In all likelihood you would not. Purchasing premade cabinets makes more sense. You would have needed the skills, time and the tools to do the same job. Making a decision to use someone else's pre-built cabinets, or in this case someone else's DNA matrix tools, is no different."

He walked back to where Elizabeth was sitting, looking at him. He then looked at the jury from the far side of the room as he said, "Do you know how many vacations she has taken in the last thirty years? Do you know how big a mansion she lives in? Do you wonder what kind of a fancy car she drives? Now remember, she is one of the top researchers in the world when it comes to genetic science. If she had chosen to go to work for some private lab, she could have earned a fortune. She chose to stay at the university. Her lab has already produced some major medical treatments that are today helping people return to their normal lives."

He paused and looked at each of the jurors, "Some of you, or perhaps a family member, is alive today because her lab provided new medical treatments. The prosecution would have you believe she lives high on the hog. The truth is, she has never once taken a vacation in the entire time she has been at the lab. Her unused vacation days number in the hundreds. Her house, it's the same one she moved into twenty five years ago. A simple ranch style house that needs a coat of paint and a new roof. And what about her car? Well, I have to tell you, she did buy a new car six years ago. She had to. Her Gremlin died and they could no longer get parts for it. She is not a woman who wants for material items. She has dedicated her life to research. Now she is on trial for embezzling funds, wire-fraud and misappropriations of funds."

He walked away from the jury for a moment. He looked toward the back row where the religious leaders were sitting. "If she had shared her discovery with the university, we in all likelihood would not be here today. But she knew the only way for the world to learn of her research and its possibilities, was to bring it to the world's attention. She realized that when the university found out, they would try and force her to not reveal the secrets of her work beyond the lab. They wanted to keep secret everything she gained through her research. So, she decided she would stay true to her life's sacrifice, she stayed the course. The goal was to make sure it would benefit everyone who needed it and not just those who could afford drug company prices."

Charlie walked back near the jury box. "You remember I told you I was unable to sleep, thinking about today? Well, I decided to turn the television on to see if there was any news. By chance there was a report about an Israeli doctor named Al-Farsi. I had heard about this story a short time ago, but the television report had picked up the complete story. The reporter went on to explain that in the Persian Gulf the doctor had found something historic. A NASA satellite a year before had detected the possible location where four rivers converged. That location is at the bottom of the Persian Gulf. His research ship drilled down through 575 feet of mud and sand looking for something lost long ago. They retrieved the sample and brought it onboard their ship and then back to Tel Aviv to be analyzed. You know what they determined? Carbon dating revealed it was between six and seven thousand years old. It was a piece of wood. They found it in a layer of deep mud which contained a type of white clay. The clay layer contains no oxygen; it did not allow the growth of bacteria. The clay protected the specimen of wood from decay. What's even more fascinating, it still has some viable DNA contained within its fibers. Doctor Al-Faris has taken his find to a genetic lab where they extracted viable DNA. The hope is it can be restored in an effort to bring back the tree. They are using similar techniques that the Koreans are using to work with mammoth DNA. And best of all, they think they can create a living tree. A tree whose genetic ancestry stood in the Garden of Eden. Who knows, it may be the tree where Eve picked the fruit. The fruit that open his eyes to the world around him. Isn't that something? If they do this thing, what have they done? They will have given us the tree from the Old Testament. Can a new Garden of Eden be far behind?

She has not. If Eden has been found and this piece of wood is from a tree that grew in that place, the words in the Bible will be deemed undeniably true. And if they are successful in using its DNA to create a tree we can touch? Wait a minute, will religion then say only God can create a tree. Or will mankind once again look upon the tree from the Garden of Eden and marvel at the beginning of creation? If he does, he will have to accept how it came to be in these modern times."

Charlie returned to his table and picked up a small book. He walked back to where the jury sat waiting for his words. He looked at them and then turned to a page in the book. They could now see it was the Bible.

Charlie began. "And God said, "Let the land produce vegetation: seed-bearing plants and trees on the land that bear fruit with seed in it, according to their various kinds." And it was so. The land produced vegetation: plants bearing seed according to their kinds and trees bearing fruit with seed in it according to their kinds. And God saw it was good. The third day."

He looked up at the jury, "God said it was 'good.' The tree that they have found was said to be, 'good' by God. How can it be anything else? God himself said it was 'good.' God in his wisdom told us as he created the world after each step it was, 'good'. But, when he created man and woman, he simply blessed them. He did not say, 'and they were good'. He knew that he had given us intelligence, the ability to think. Man is not always 'good.' And yet He allowed us to be intelligent and make choices. Religion and science are not at odds with each other, they are a way to re-verify those gifts which mankind has been given and our intelligence is one of the greatest gifts."

There was a hushed mumbling in the courtroom as some whispered among themselves as Charlie told his story. "Hearing about Doctor Al-Fairs' discovery gives me wonder, does the discovery in the Middle East affect the case you sit in judgment on today? Neither would be possible without science. Can we in one case say that we are okay with creating a tree? And on the other hand say we are not okay with creating a man?"

He walked away from the jury toward the center of the room. He stopped and turned back, raising his hand toward the ceiling. "Then it came to me. The quest to find the tree is validation of religions basic beliefs. But remember, they needed science to prove what they had found. Radiocarbon dating… extracting DNA from a piece of wood… locating the site from space. This is the science of man. Still, some will say it is a miracle by the hand of God."

With a questioning look he asked, "A Biblical miracle that only science could create? Just think, perhaps in twenty years you will be able to travel to the Middle East and for a fee, see the Eden Tree. Those who financed the research will in all likelihood patent the DNA and the created tree. They will profit from the genetic scientific discovery from the Biblical birthplace of man. Before you today is Doctor Mathenson. She will soon await your judgment. A judgment that may help to save the future of man from pain and disease."

Charlie walked over to the defense table and touched the newspaper Dobbs had used to tell the jury about its headlines. As he touched it he looked at the jury. "This is not a trial about evolution or some theory of man rising from an amoeba or even a monkey. It is about man and his frailties when it comes to diseases. It is about Doctor Mathenson and her lifelong quest to find the answers to life. She has not claimed to be God or even a great person, just someone who made a difference. In Israel they are celebrating the evidence of God's Eden. We should be celebrating and believe the creation of man is by the hand of God. That same God gave us the ability to help man along the way in this world of ours. Elizabeth may have acted in poor judgment with the financial matters of the university, but she did not show malice of forethought or personal gain. I ask you to judge her on the entirety of what you have heard over the last week. She went looking for the truth and along the way she found more than she expected. Should she be found guilty because she did her job and more?" Charlie paused and looked at each of the jurors before turning and walking back to his seat. The room was hushed.

Judge Thompson folded his glasses and laid them on his desk. Then he looked at the jury, "Ladies and gentleman, you have heard the evidence presented in this room. I would tell you that only the information presented during witness statements and physical evidence entered can be used in your consideration of the defendant's guilt or innocence. The bailiff will take you to the jury room so you may begin your deliberations." With that the bailiff motioned for the jury to follow him from the room. Judge Thompson recessed the court until a verdict is reached.

Elizabeth looked at Charlie. "What now? Do we wait here or do we return to the hotel?"

Charlie closed his briefcase and laid it flat on the desk. He then looked at Elizabeth, "We wait. The jury may be out an hour or not even finished today. Hard to tell. I think most of them have made their decisions already, but a few may still be trying in their minds to justify your actions. Dobbs closing changed everything. I thought by his presentation during the witness testimony he would stay pretty much with the financial evidence. But he clearly wanted to join the financial questions in with the ethical subject. I had to think on my feet about my closing after hearing his summation."

Charlie looked around the courtroom as some filed out and others sat waiting. "We may as well make our way back to the defense counsel conference room. We'll send downstairs for coffee. All we can do at this point is get comfortable and wait.

Chapter 28
The Verdict

Morning passed and the clock on the wall showed the afternoon came and soon passed. Charlie and Downing took a walk around the block to stretch their legs. They talked not of the trial, but what Downing was planning to do after his first case. Charlie gave him some advice and told him he believed he was going to be an excellent trial lawyer, if that was his choosing. Downing said at least for a while, personal wills and deed searches looked pretty good.

The sun was just touching the hills in the west when Charlie got the call, the jury was back. They had taken eight hours for their deliberations. Clearly there had been some disagreement among the jurors. Charlie hoped they were able to resolve them into a clear judgment. No hung jury. He felt the call of Smoke Hole Canyon calling him and he wanted to answer that call.

A half hour later everyone was back in their places. Judge Thompson motioned for the bailiff to bring in the jurors. As they filed in they looked not toward Elizabeth but toward the judge.

Thompson asked, "Madame Foreman, has the jury reached a verdict? A short lady in a flowered print blouse stood with a paper in her hand. "Yes, your honor, we have."

The judge asked the bailiff to bring it to him, so he could read it. He adjusted his glasses as he read the paper. He then handed it back to the bailiff who returned it to the lady.

Judge Thompson looked toward the defense table. They stood not knowing what he was going to say. "Doctor Mathenson, a jury of your peers has reached a verdict in the charges against you."

He turned toward the jury box. "The foreman may now read the verdict."

The lady's glasses hung around her neck like those of a school teacher. She placed them on to better read the paper. "As to the count one, embezzlement of funds outside the guidelines of the project goals…we find the defendant….not guilty."

The sounds of voices whispering filled the courtroom after hearing the first charge announced. Judge Thompson struck his gavel and said, "If you want to hear the rest of them from inside this courtroom, I want to hear no more expressions of opinions."

He nodded to the foremen to continue her reading, "Count two of the charge of embezzlement of funds not within the discretion of the project goals….we find the defendant…not guilty. As to the charge of wire-fraud and misrepresenting solicitation of funds….we find the defendant….not guilty. As to the count of misuse of project funds without proper over sight as required by federal grant guidelines….we find the defendant ….guilty."

Again the courtroom was abuzz with whispered voices. Two reporters bolted for the doors trying to be first with the verdicts outside the courtroom. A moment later loud voices could be heard outside the doors.

Thompson gaveled the court back into order. He then addressed the jury, "Do you all agree with the verdicts?"

The group all shook their heads in the affirmative.

Thompson said, "I thank you for your service."

He then turned to Elizabeth, "Doctor Mathenson, you have been found guilty of misuse of project funds without proper over sight as required by federal grant guidelines. I am trying to decide whether or not to pass sentencing at this time or give it more consideration. Before I make that decision, do you have anything to say to the court?"

Elizabeth paused for a moment. "Yes, your honor, I have a few words, if I may."

Thompson nodded his approval and sat back in his chair, "I wish to apologize to my colleagues for my actions. My intent was to further our work and project goals. I freely admit I took no interest in the question of the money and how it was administered. As the head of the lab, I understand it was my job to help manage the money as much as it was to work on the project's research. As to my action with the embryos and their being transferred to long term storage for the future, I have no regrets. I believe if my actions preserve that science and spark conversation about the future work of genetics around the world, I have accomplished my personal goal. I thank you for allowing me to address the court, your honor."

Judge Thompson removed his glasses and rubbed his eyes. He then looked at Dobbs and then at Charlie. Finally he returned his attention to Elizabeth. "Doctor Elizabeth Mathenson, this court has found you guilty.. The charge carries a sentence of one to five years in confinement, but I feel justice would better be served if I were to place you on probation for two years and sentence you to one thousand hours of service in a cancer ward at a hospital. Perhaps that experience will give you an idea of the importance of your work. That is the decision of this court." With that he gaveled the court into adjournment.

Elizabeth turned to Charlie, "Does that mean I get my job back?"

Charlie hesitated, "I am sorry Elizabeth, I am afraid it does not. He has sentenced you on the charges brought against you. Your dismissal from the university is another matter completely."

Downing placed his hand on her shoulder as tears ran down her face at hearing Charlie's explanation. "Don't worry doctor, I'll file a brief in a couple of days with the university board asking for complete reinstatement. I think they will want you back, especially if you are working days in the cancer wing of the hospital."

Jake, Angus and Tucker all shook Charlie and Downing's hands. Each also hugged Elizabeth.

About then Elizabeth saw her son and his wife making their way toward her. She could not help but smile as she greeted them. Her daughter in-law held a small baby only a few weeks old. Elizabeth made over him as she touched his face. She took the baby in her arms and walked toward Charlie. The small baby held Elizabeth's finger tightly in his hand. She looked up at Charlie and said in a very gentle voice, "Charlie, I want you to meet my grandson…Adam."

Charlie knew at that moment what she was telling him. He remembered that the original inventory of frozen eggs contained one-hundred and twenty-five. The inventory sent to England contained one-hundred and twenty-four. He had wondered about that discrepancy, but just figured it was an error when recording the information.

Charlie smiled at her and said, "Life is precious no matter how it comes to be. I miss my son, I believe he would have liked to have known your grandson. I hope he is everything you have hoped for. But, I think the world is not ready for the truth. At least not yet. Good luck to you and your family."

Chapter 29

Return to the Smoke Hole

Charlie returned to his home and for a few days sat on the porch drinking sweet tea. He enjoyed watching the neighbors mow their lawns. He knew they mumbled about the fact his lawn needed mowing. He thought about firing the kid next door for failure to honor his contract with Charlie, but he decided to let it grow long and piss off the neighbors.

Jake returned to his home on the lake and placed more no trespassing signs around his land. He texted Charlie on his burner phone that the crappie were biting just off his dock. Charlie figured he still wanted to make sure no one was listening to their conversations. He had also returned to the senior center and went on teaching old ladies how to game online. He told Charlie they may be old, but they could be vicious in combat games and took no prisoners

A few days after returning to Elkins, Charlie got a visitor, it was Clay Gordon. Clay told Charlie before he began, "This conversation is to be off the record." Charlie understood. Clay went on to explain that his government was going to drop the request for the return of the specimens. He also explained that some in his government felt preparations should be made for the future. He explained that much of the confusion came from those who believed their country should isolate itself from the outside world. He went on and asked if Charlie was going to return to speak to his classes this fall. Charlie told him he was looking forward to it. In fact he told Clay he would like to return on a regular schedule to speak to his classes. Clay very much agreed. They never spoke of Clay's position with his government. Clay also was good to his word, helping Jud and Samantha get into law school. He somehow got them scholarships from the University of Jerusalem to continue their studies of international law. He also told Charlie that if he ever wanted to see the piece of wood they found at the bottom of the Persian Gulf, he could arrange it. Charlie declined his offer.

James Downing filed a petition for reinstatement with the university board of review for Elizabeth. They rejected the petition. They felt when she removed the embryos and the documentation from the lab, she violated their trust. They did, however, allow her to retire and returned her tenure. Elisabeth would later accept a position at Stonemore with Doctor Whitechurch. She helped Angus with his DNA Ark Project.

James Downing turned down several offers from large law firms wanting him to be part of their partnerships. He chose to stay in his office and practice real law as he saw it.

Harrison Dobbs never got his Supreme Court appointment. But he was asked to head up a commission to find answers to the legal questions that genetic science was presenting to the courts.

Reverend Jedidiah Cain used the trial to promote his beliefs that Satan was at work in the world and his words from God were the only salvation for mankind. Later it was reported that his church was being investigated for tax evasion. Under his guidance the church had purchased several property holdings in the Bahamas. Two houses and a large yacht. They claimed they were the church's holdings and exempt from taxes. Jedidiah denied any wrong doing. He claimed God wanted him to have a place of refuge from the evil in the world. The IRS did not see it that way.

As to Elizabeth's grandson, Adam, Charlie had heard from Downing the boy was growing quickly for his age. He seemed to have an awareness of the world beyond his youth. He began crawling at five months and speaking short phrases at one year. His parents have high hopes their son will grow into a fine young man.

As the leaves of summer faded the colors of fall returned to the Smoke Hole Canyon. Charlie had returned to try and catch The General once again. He felt by the pain in his knees, that it would be the last time he could venture into the canyon on his own. But he was at peace with that knowledge. It had been a place where he had found solace from the world since he was a boy.

On a cool crisp fall morning he sat looking into the cold rushing waters of the canyon. He knew his old adversary was waiting in the shadow of the rocks. Charlie savored the moment as he remembered other years he came here with his dad. He thought of the times he had anticipated bringing Timmy here and teaching him the tricks of catching fish. All those things were now behind him in his life. He decided to take Nancy's advice and write a book about fishing using the stories told to him by his dad. He included some of his own stories of fish caught and the ones that got away. Maybe some other kid would enjoy hearing about the old days fishing the Smoke Hole Canyon. The last year of his life had changed him. He somehow had a bigger picture of how the world works. He also felt that one day he would be with Rose and Timmy once again.

He waited until the sun was just right, shining on the fast waters just below the riffles near the far rocks. He cast his line to a place just above the sunlit area. The surface current began to move his fly down through the shadowed swift water.

Suddenly, his line went tight and he felt the tug of a large fish. He hoped it would be The General coming out to challenge him one last time. As he maneuvered the fish from deep water, he saw a bright flash of gold. He knew it was not The General, but a golden trout.

Slowly Charlie maneuvered the struggling fish toward the shallows where he would be in control. Finally, he had him where he wanted him. He reached down and gently cradled the golden fish in his hand. The fish's body felt cold from the water's embrace. As he held his catch he could feel the fish's heart rushing in fear. Its jaws opened and closed, trying to bring oxygen to its rapidly beating heart. Charlie realized the fish was a creation of God, even if man had a hand in helping it to be part of this world. He now realized that was the plan all along. Charlie knew he had no reason to sit in judgment of the fish and how it came to be. Life and the wonderment of it should be enjoyed. Charlie now knew the answer to Jedidiah's question, "Did he believe in God." He smiled and spoke gently to the fish, as if to ease its fears as he removed the hook. He then lowered the fish into the cold waters and watched him swim away. But just as it was almost gone, it turned and circled, seeming to look back at Charlie. Maybe even a fish can wonder about mankind and his future. The fish disappeared into the deep dark waters of the Smoke Hole Canyon. Charlie Morgan, for the first time in many years, felt at peace with the world and his faith.

The End

Made in the USA
Monee, IL
04 November 2026

67376991R00262